Body

BODY

Don Johnson

Beacon Press   Boston

Grateful acknowledgment is given to Pantheon Books for permission to reprint European head bindings from *The Body Reader,* edited by Ted Polhemus, and to the Rolf Institute for permission to reprint its logo; copyright ©1979 Rolf Institute. The tortoise and scorpion postures are reprinted by permission of Schocken Books Inc. from *Light on Yoga* by B. K. S. Iyengar ©George Allen & Unwin (Publishers) Ltd., 1966, 1968, 1976. Zazen is reprinted by permission of John Weatherhill from Koji Sato's *The Zen Life* ©1966, photo by Sato Kuzunishi. Grateful acknowledgment is given to Ira Friedlander for permission to reprint the photo of the dervishes. *Virgin and Child* courtesy, Museum of Fine Arts, Boston. William Francis Warden Fund. The normal body in Western medicine is reprinted by permission of the Anatomical Chart Company. The nineteenth-century orthopedic devices, the Milwaukee brace, and the chiropractic ideal spine are reprinted by permission of the W. B. Saunders Company. The Alexander spine is from *The Alexander Technique,* by Wilfred Barlow, copyright ©1973 by Wilfred Barlow. Reprinted by permission of Alfred A. Knopf Inc. and by permission of Victor Gollancz Ltd from *The Alexander Principle* by Wilfred Barlow. The profile of the spinal column from Ida Rolf's *The Integration of Human Structures* is reprinted by permission of John Lodge. The ideal computer operator is reprinted by permission of Henry Dreyfuss Associates and M.I.T. Press.

Beacon Press books are published under the auspices of the Unitarian Universalist Association of Congregations in North America, 25 Beacon Street, Boston, Massachusetts 02108 Published simultaneously in Canada by Fitzhenry & Whiteside Limited, Toronto

Printed in the United States of America

*BF
161
.J59
1983*

(hardcover) 9 8 7 6 5 4 3 2 1
(paperback) 9 8 7 6 5 4 3 2 1

Library of Congress Cataloging in Publication Data

Johnson, Don, 1934–
  Body.

  Bibliography: p. 221.
  Includes index.
  1. Mind and body. 2. Mind and body — Social aspects.
3. Mental health. 4. Health. I. Title
BF161.J59 1983   128'.2   83-70653
ISBN 0-8070-2900-9   ISBN 0-8070-2901-7 (pbk.)

For my father,
a protestant

CONTENTS

Body

CHAPTER ONE

SHRINKING BEFORE AUTHORITIES

This book is about healing fractures in our personal and social bodies that cripple our ability to take firm stands and move freely.

I recently dreamed that I broke my leg while living with two orthopedists. One, older, with short hair, and dressed in a white smock, prescribed putting my leg in a cast and bed rest for six weeks. He told me that unless I kept the leg still its tissues would not heal properly. The other, younger, long-haired, and dressed more casually, objected to both the cast and the bed rest. Movement, he argued, would heal me faster and with less residual scar tissue. He prescribed a continuation of my ordinary activities supplemented by a vigorous exercise routine. I explained to the doctors in great detail what you are about to read, giving reasons that I would have to experiment with different ways of healing myself and see what happened. I told them I would appreciate their telling me everything they knew about fractures and giving me opinions about my experiments. But I would have to be the ultimate judge of the appropriate course of action.

The dream summarized my peculiar explorations in how to mend our fragmented social world. Early in life I learned to disconnect my philosophy, politics, and spirituality from my body. In that weakened state of alienation, lacking a sense of my own resources for making judgments and decisions, I had to rely heavily on publicly recognized authorities. I spent nearly half a century trying to find the right people with the right prescriptions for successful living, ranging from the most traditional to the avant-garde. I had learned to discount my own resources for finding my way through life and for evaluating the advice of others. I finally realized that a basic ingredient for healing fractures within our personal experience is learning how to reconnect our more abstract attitudes with our sensual experiences. This book examines ways in which those connections have been broken and avenues of return.

The dream, which was about who is in charge of mending broken limbs, also expressed the bodily focus of my search. What kinds of knowledge count in the shaping of our muscles and neural pathways? What is the value of my own experience of my body in relation to scientific knowledge possessed by experts?

My attitude toward the doctors in the dream reflected my family's ambivalent attitudes toward experts. All of my ancestors as far back as I know were artisans and farmers from Scandinavia, England, and Ireland. Many came to the Sacramento Valley during the final years of the Gold Rush, wanting to live in a world freed of the pressures brought to bear on people by monarchs, bishops, and plutocrats.

I was born in the final years of the Great Depression, and was raised in an atmosphere severely critical of greedy politicians, industrialists, aristocrats, and pretentious intellectuals. But at the same time I was raised a Roman Catholic, a member of a royalist Mediterranean religion which never grafted too easily onto my Viking-Celtic roots. Catholicism was my mother's religion. My father signed me over to it before I was even born, in a contract demanded of Protestants who married Catholics.

Early in life I received a bewildering number of nostrums, some for my physical body, others for my soul. Once, during the course of psychotherapy, I listed all the bodily prescriptions I remembered learning as a child and came up with nearly two hundred. They included such specifics as "If you sit in a draft, you'll get a cold"; "If you climb trees with your friends, you'll break an arm or a leg"; "If you go hunting, you're liable to get shot by accident"; "If you read a book lying in your bed, you'll get eyestrain." Some had to do with fitting into society: "Stand up straight if you want to be attractive"; "Shake hands firmly or people will think you're a weakling." The most important had to do with salvation: "If you touch your penis with pleasure, or enjoy looking at your naked body in a mirror, you're committing a mortal sin, worthy of eternal damnation."

The prescriptions came from a lucidly defined hierarchy. The basic diagnostic division was between Catholics and non-Catholics. I'd best not listen to the latter, nor associate with them, even though my father was one. In later years I learned more subtle diagnoses descending from the mildly diseased Protestants (who believe in Christ) to Jews and Muslims (who don't believe in Christ but believe in his Father) to Unitarians, Hindus, and Buddhists (who seem to believe something) to the crippled agnostics (who at least admit they don't know) to the terminally ill atheists.

My mother's and my immediate superiors were Monsignor Renwald, pastor of the Sacramento Cathedral, and the nuns whom he assigned to teach me catechism. Right next to Monsignor in my childhood perceptions was Doctor Savarin, our family doctor, the old-fashioned kind who made home visits and actually touched me to comfort me. I especially liked his prescriptions for my debilitating asthma, like the opiated cough syrup Sedatol, which put me into reveries about exotic worlds far removed from the Sacramento Valley. All of us, including Doctor Savarin, were subjects of Bishop Armstrong, previously a farmer from Tacoma, who lived only two blocks away. He himself was answerable to the aristocratic Pope Pius XII, the chief orthopedist of my soul.

The Christian Brothers, my high-school teachers, were inferior to priests but above nuns because they had the genital equipment to become priests. Jesuits, who taught me in college, were a cut above regular priests because they were subject only to the Pope.

Sister Felicia, the Franciscan nun who taught me catechism as a young boy, said that the hierarchy extended into the invisible world. Hovering just above the heads of all human beings, no matter how important, were the angels; each of us had his or her personal guardian. Above them were archangels, who guarded presidents and kings; principalities, who took charge of cities; powers, who watched over whole nations; and finally the seraphim and cherubim, who surrounded the throne of the trinitarian God. Down below, of course, were the myriad devils and Lucifer, the fallen archangel. The life I saw and felt about me was only a screen concealing the true drama in which these invisible forces were warring over my soul.

In 1956, at age twenty-two, I joined a religious order of men called the Society of Jesus — the Jesuits — founded by the Basque Ignatius of Loyola in 1564. I tried several new prescriptions for healthy living. My vow of chastity bound me to behave, in the words of Ignatius, "like the angels," not even entertaining a sexual thought or impulse. My vow of obedience enjoined me, again in his words, to "bend my mind into conformity with that of my superior," behaving like an old man's staff without a will of my own. We novices were given instructions about how to carry our bodies erectly, where to place our eyes when dealing with superiors, and how to insure that our hands never drifted near our genitals.

But the Jesuits were more than an authoritarian, papal version of the CIA. They were also a revolutionary band of intellectuals committed to exploring the furthest reaches of human thought. In my early years within that group I was encouraged to immerse myself in contemporary philosophy and psychology. I encountered another class of "orthopedists," loosely associated with a fledgling "human potential movement." They disputed the old prescriptions and argued for new ones, like those of the second orthopedist in my dream. In essence, they said I had given up my body in favor of my mind. People such as Carl Rogers and Rollo May encouraged me to express my feelings and follow my own impulses. Alexander Lowen, the founder of bioenergetics, argued that if I didn't express my anger, I might get cancer or crippling arthritis. Many people, both Jesuits and human-potential experts, prescribed psychedelic drugs instead of the

Sedatol and antihistamines that I had taken regularly until then. The elderly Dr. Ida Rolf said that if my third cervical vertebra did not move slightly back and the space between my eleventh and twelfth ribs increase about one-eighth of an inch, I could not even understand something as basic as the science of anatomy. Movement teachers argued that if I didn't let my weight fall forward at just the right angle when I walked, I might never get rid of my back pain.

For nearly ten years I hovered between the advice of the traditional and the avant-garde. But in 1970 I decided in favor of the latter and left the Jesuits to join the small band of therapists studying with Dr. Rolf.[1] I was convinced that society's ills were the result of renouncing the body in favor of the mind, so I retired to the quiet town of Santa Fe, New Mexico, to pursue knowledge of the body.

During the next ten years, which I spent working with people, several cases and personal experiences made me think I had not accurately assessed the importance of the body in my own life or in society. A typical client, for example, was a fifty-five-year-old physicist who had held a respected position as a researcher at the government laboratory at Los Alamos for some twenty-five years. Let us call him Walter. He was six feet, three inches tall, with a muscular body well toned by frequent backpacking trips and rock-climbing expeditions. He came to me complaining of chronic headaches: a few times each week, pain would begin in the back of his neck and radiate up over his head. I worked with him for several weeks, manipulating his body and asking him to examine the history of his bodily life. One day he arrived at my office very excited about an insight he had had while riding in the elevator with the lab director. He said he had been feeling fine that morning, but as soon as his boss got into the elevator and began speaking to him, Walter felt his body shrinking like a little boy's and his neck beginning to hurt. He suddenly realized that the same thing happened every time he saw his boss, which was several times a day. More significant, he remembered that he had always acted that way around his father. Walter's headaches gradually subsided from that day on.

What particularly struck me was that Walter's body was in excellent condition. He was strong and reasonably well balanced from a structural viewpoint, having no extreme curves or tilts. He seemed to have an active and satisfying sexual life with his wife of many years. His diet was moderate. It could not be accurate to say that Walter had sacrificed his body for his mind.

Another fact that seemed peculiar was that although Walter had engaged in psychotherapy for some years, he still repeated the muscular behavior he had used as a child who was afraid of his father. Moreover, his deferential behavior was directed toward a boss who was considerably younger, had less experience than Walter in physics, and had hardly any administrative experience.

Although Walter had cared for his body, he had not paid attention to it as a source of significant knowledge. He had not learned to relate such banal things as headaches to his submissiveness toward authorities.

Similarly with my own life. It could hardly be said that I had ignored my body in favor of my mind — I had always given an inordinate amount of attention to the physical and spiritual health of my body, had exercised regularly and kept in good physical shape. Changing the old for the new-age wing of "orthopedists," with their more sophisticated prescriptions for my body, seemed actually to have had little effect. Chronic pains persisted, and I seemed no more capable of intimacy with the people I loved. I noticed that I too was still shrinking in the face of authorities, even though I was no longer subject to a vow of obedience.

I began to reassess my acceptance of the popular myth that our culture has ignored the body in favor of the mind. At face value, I realized, that statement cannot be true. Never before has the body received so much attention. Health is a national priority. Exercise has become a part of everyday life: parks are filled with joggers; health spas, athletic clubs, and dancercise classes thrive. Stress-reduction techniques are often incorporated into programs that businesses provide for their employees. Massage and body therapies, once considered fronts for prostitution or quackery, are now prescribed by physicians and paid for bý insurance. Methods to reduce the traumas of pregnancy and birth have been disseminated. The American public spends $287 billion a year on health care — more than the entire defense budget, the yearly U.S. sales of foreign and domestic cars, and the profits of forty-one major oil companies.[2]

Diet is a congressional concern, and the government publishes recommendations about which foods to eat and to avoid. The success of health-food stores has provoked supermarkets to devote large sections of vitamins and health aids. The emphasis on natural foods and whole grains, once found only in specialized vegetarian restaurants, now characterizes haute cuisine.

Vast numbers of people, ranging from singles in their twenties to retirees in their seventies, exhibit well tanned and toned bodies, energetic and ready to move. Cosmetics is one of our major industries. Aesthetic surgery, formerly the preserve of public figures, is now widely used by men as well as women.

Sexuality has been liberated from its former constraints. Sex education, methods of birth control, and sex therapy are easily available, supported by the government and insurance companies. Nudity, homosexuality, and extra- and premarital sex have become more acceptable. One no longer has to slip furtively to select cigar stores to find the cheaply printed magazines that reveal what genitals look like; any magazine shop carries a variety of slick journals with high-quality photographs of every imaginable form of male and female sexuality. Adult motels and elegant clubs allow people with enough money to pursue every sexual fantasy. Dark, smelly, X-rated movie houses have been supplemented with comfortable, ordinary theaters that screen artfully done sex films and with home videos. Orgasms, once the secret, rapid pleasure confined mostly to males, have become a national recreation, with a seeming increase both in frequency and quality.

The chain bookstores tell the story. Each has major

sections on the body: health, nutrition, sex, exercise, beauty. As I write this, a third of the fifteen nonfiction titles on the *New York Times* best-seller list are about our physical nature. But this public enshrining of physicality conceals some illusions; care of the body has replaced religion as our opiate, dulling our sensibilities to the degradation of physical life on the planet. It turns out that the body that is supported by government programs, insurance companies, and corporate funding is the body that is useful for our dominant social goals. Those goals are economic and short-term. In the long run they endanger the well-being of the human species, a community of bodies nourished by a physical environment.

As white middle-class Americans perfect their bodies, the skies become murkier, lowering water tables show increasing signs of chemical poisoning, and the northern forests turn brown under acid rain. Studies begin to reveal the long-term hazards of low-level radiation and toxic chemicals for the health of masses of workers. On a planetary level, proportionately more people are dying of starvation than ever before in recorded history. Twenty-five percent of the world's population is living in absolute poverty, and the number is rapidly increasing. Nuclear weaponry puts into question the continued existence of physical life itself.

Moreover, the popular emphasis on the body does not necessarily lead to liberation in the sense of achieving more power to accomplish basic human goals. The body that is the object of so much attention is only a partial body; it is not the body as a source of intelligent decision-making. The many sensory and emotional resources we possess for finding our way together through a sensual world are impoverished by the way our physical selves are molded in the culture. The successful products of the various shaping agencies of our society are adults, like Walter and myself, who feel ignorant in the face of the supposedly complex knowledge they need to make the most significant decisions about their lives, and helpless in altering national policy. The outcome of our socialization is a citizenry that becomes flaccid when standing in front of authorities.

Wilhelm Reich called that flaccidity our "physiological incapacity for freedom." "All freedom fighters," he wrote, "have made this miscalculation: The social incapacity for freedom is sexual-physiologically anchored in the human organism. It follows from this that the overcoming of the physiological incapacity for freedom is one of the most important basic preconditions of every genuine fight for freedom."[3]

This book examines different aspects of that physiological impotence:

how the supple and exploratory bodies of infants turn into the machinelike bodies of adults like Walter, with predictable stances and ways of seeing things;

how Walter and others of us learn to give up our sense of physical vitality and experiential authority in the face of those whose only claim to superiority is a higher salary and an appointment from above;

why tensions and emotional repulsions constantly break up attempts to mobilize our community resources;

how we learn to trust politicians, physicians, and religious functionaries more than our sensual experience and that of our friends.

These why's and how's have psychological and sociological answers, and much has been said about them. But I am interested in the fleshy dimensions of those answers.

Pause

Periodically I will invite you to halt the onward movements of thought and to return to yourself, reflecting on such things as your muscular tensions, the quality of your sensations, your memories, and your typical postures. In addition to the pauses I offer in the book, I suggest you use this technique to recover a sense of the power of your experience in your ordinary life. When you are engaged in an intense telephone conversation, you might stop for a fleeting moment to notice what's happening in your body. While making love, occasionally check to see if you're holding back or moving mechanically. Sometimes you might want to make the pause more defined. In the midst of a frantic day you may find it useful to spend five minutes resting in your chair or lying on the floor with your eyes closed. When you're jogging, a two-minute pause might help you recover a sense of ease that dissipated in the rush of anxious determination. These interruptions of ordinary activity are a basic means of reconnecting fractured parts of our experience.

In this first interlude, I invite you to make the following reflections:

Think of the earliest years of your life you can remember. Visualize your house or apartment, its kitchen, bedrooms, living room. Remember what your parents looked like then. Recall a scene when your father was asserting his authority. What did he look like? What gestures did he use? Postures? How did your mother look when he did this? Can you recall how you felt in your body?

Recall a scene when your mother was asserting her authority. How did all your bodies behave when that happened?

Recall an incident when your mother and father were in the presence of people they regarded as authorities (a physician, minister, priest, boss, an "important" person). Did their bodies change? Their facial expressions? Gestures? How did their behavior make you feel?

When you're now confronted by people you consider authorities, do you adopt distinctive postures? Does your body have characteristic feelings?

I've titled the explorations in these pages *Body* because at one point in my examination of these issues I discovered that our disregard for the sources of wisdom in our flesh are accurately reflected in the history of language. The word *body* is derived from the Anglo-Saxon *bodig*, the Old High German *botah*, and the German *bottich*, which means a cask, a brewing tub, or a vat.[4] Throughout its twelve-hundred-year history, the word's primary meaning has been "the whole material organism viewed as an organic entity"; "the material being of man as the sign and tangible part of his individuality, taken for the whole"; the "person," as in the 1549 *Book of Common Prayer* ("With this ring I thee wed...with my body I thee worship"). Or, as John Locke wrote: "One angry body decomposes the whole company."

Body has commonly been used interchangeably with *person* to refer to a human being of either sex, as in "If a body meet a body comin' through the rye," and has been combined with *any, every, no,* and *some.* The word was used for centuries to refer to the vessel used by alchemists for transmuting their elements, since the human body's capacity to change food into wisdom was the model for the alchemists' goal of deriving pure gold from base metals. Until the last century, *body* was used rarely of the dead body, and then only as a euphemism or abbreviation.

During the past two-hundred years, however, the lively tradition of the word began to be infected by a set of meanings associated with the Latin word *corpus* (plural: *corpora;* from the Sanskrit *krp*), which has dominated the intellectual history of Europe. The previously wide reference of *body* to the whole person became reduced to a small part of what we call human.

The meanings associated with *corpus* reflect the nature of the compound fracture examined in this book. The fundamental break is between what we call "mind" and "body." This division radiates into breaks between, for example, ordinary people and experts, man and woman, us and them (the enemies), science and common sense.

Corpus is one of those words that is defined correlatively, like *right, left, inside,* and *outside.* If you look up such a word in the dictionary, you will find it defined in relation to its complement. The partner of *corpus* is the soul. *Corpus* is that in man (originally, specifically in man, because slaves and women were thought not to have souls) that is distinct from, or in opposition to, the soul. *Corpus* means "the body of man considered as the seat of strength, virility, sexual activity, physiological needs and desires; in sum, the basis of the grosser elements in man. It is commonly used to refer to the dead body."[5]

This is the only Latin word for the body, and the only words in the Romance languages are derived from *corpus:* *corpo* (Italian), *cuerpo* (Spanish), *corps* (French). *Corpus* was transformed into the German *körper* and the English *corse,* eventually *corpse.*

The religious and philosophical tradition underlying the history of *corpus,* articulated by such people as Plato, St. Augustine, and Descartes, placed minimal value on bodily existence. *Corpus* is what exists to receive the soul at birth, what's left after the soul departs. It is both kit and tools for the soul; the soul's Ferrari, Dodge pickup, or rattletrap; its sometimes fashionable, sometimes dowdy clothing. Justice, love, and depth belong to the soul, whereas *corpus* gets lust, hostility, and weight. With *corpora* we can do no more than penetrate each other with small parts of ourselves; it is through souls, according to this way of thinking, that we can become truly one.

Corpus was the only word available in sixteenth-century European universities when Vesalius created the science of anatomy, a science of cutting up (anatome) cadavers. Corpses of criminals thrown on garbage heaps outside of Brussels were stolen away at night and cut up by candlelight in hidden cellars. The muscles, bones, and organs depicted in anatomy books — the foundation of modern medicine — are those of dead outlaws cut up in secret.

Western rationality is like the anatomist's scalpel, the process of scientific understanding like dissection. The objects of understanding are thought to be inanimate, purified of so-called subjective qualities more appropriate to the souls of the observers. For Western culture, life in the body has been haunted by images of death: *corpora* of Jesus on the crucifixes in every European village; corpses of soldiers (Caesar's, Attila's, Charlemagne's, Patton's) lying in graves beneath those crosses; *corpora* of ordinary men and women dead to feeling after centuries of repression and decades of television; anaesthetized *corpora* lying in surgery rooms; the fragile *corpora* of infants in places like Somalia and Cambodia, waiting to die while their food is eaten or destroyed by armies clashing for the sake of ideas conceived in Afro-baroque palaces or in the boardrooms of Western metropolitan skyscrapers.

Corpus is a thing which is also thought of, and treated as, a machine: a collection of particles (*corpuscles*) arranged in space, interacting according to the laws of physics. In contrast, *body* evokes images of tending a pot brimming over with heady mead or of oaken casks in which Armagnac is aging. It's an appropriate word for the ruddy bodies in a Breughel painting of a country fair, delighting in food, ready for love.

The experience signified by *bottich* is never too far removed from our consciousness. Many physicians have been initiating movements to educate people about their own resources for healing themselves, and have also been resisting pressures to integrate medicine into the technological ideology that reduces the body to a machine. Educators argue for the need to redesign our schools so that children can learn to expand their capacities for sensation and movement. Physical-fitness experts, ecologists, and community organizers are among the many who argue for a need to return to a more organic vision of social life.

In our personal lives, there are moments when all of us feel the unique qualities of our life's experience. We glimpse the brightness in our children and marvel at their creativity. In a moment of crisis we feel the wisdom of a friend or lover who is at hand to help us. At these times we know that we too are at least as sensible as anyone who might have more authority. But these impulses toward appreciating our native wisdom are bound by ropes knotted in our flesh by centuries of corporeal training.

The tending of *corpora* requires mechanics and embalmers; a body has different needs. Conceiving of the human organism as a brewing vat implies that its genius — its wisdom, justice, intelligence, love — is to be won through patient, cooperative work with tangible reality. Producing unique spirits requires a refined sense of smell and taste, the patience to respect natural processes of fermentation, and a sense of the right grains, soils, and water. A community of bodies, unlike the random collection of reactive corpuscles, requires a more enlightened politics which recognizes the unity of all bodies, human and nonhuman, and the insanity of imperiling these substantial bodies for the sake of ghostly ideologies.

But we are gripped by ideologies. The dominant values of our culture insinuate their ways into our neuromuscular responses, shaping our perceptions of the world. Altering the morbid dynamics of our culture requires us to loosen their hold on our flesh.

CHAPTER TWO

A LOSS OF SENSE

The ways in which dominant cultural values get a hold on our muscles and nerves, disconnecting us from our own experience, were brought home to me in working with another client. Let us call him Charles. Charles was sixty-five years old and had recently retired from a lifetime of traveling the world as a CIA agent. He belonged to an upper-class New England family, had graduated from Exeter and Harvard, and played polo. Tall, elegant in his movement, and well versed in literature and science, he loved good food, drink, and sex. He came to me because he felt he was at a crossroads in his life, that he was doing nothing of significance. He longed to return to a small country in Africa where he had made friends with leaders in the new government, and he could envision using his considerable expertise to help them organize their venture. But he felt impotent to take the steps necessary to follow this or other paths. He had an intuition that direct work with his body would carry him through the impasse.

As I worked with him over a period of several weeks, I was struck by the contrast between his obvious intelligence and wit and his indecisiveness and passivity. I was intrigued by the way in which he responded to my questions about what he felt in his body. When I asked him, for example, "Do you now feel more weight on one foot than the other?" "After this session of work, do you feel any different from before?" "How would you describe your physical energy right now?" he always replied in the same manner." Well, after all, how

15

can one really say?" "How can one be sure?" "Perhaps my weight is more on my right foot than my left, but can one really know that one is not being deceived?"

Charles's responses did not have the quality those same words would have in the mouth of a college sophomore who has just discovered Descartes. They revealed a childlike confusion and weakness, strangely inappropriate for such a mature person. Puzzled, I continued to explore why he was unwilling to admit even the simplest of perceptions, such as the feel of a breeze or of the texture of my carpet on the soles of his feet.

One day he got up off my therapy table after I had been manipulating the muscles of his legs and exclaimed, "What a strange memory! I must have been four or five years old. I had just listened to a record, a piece by Ravel, if I remember correctly. My nanny had played it for me. I liked it very much. I ran downstairs where my mother was standing and told her how much I liked the record. 'Why, you're such a silly little child,' she said. 'That's a trashy piece of music.' I never recall telling her again how I really felt about anything." As he told me of this experience, his elegant body visibly shrank, head dropping, shoulders slumping like a sad little boy.

Here was the best and the brightest: Charles had received the finest intellectual education; his body had been trained in the most refined sports, posture training, and orthopedics; he had discriminating taste. Yet he reported a deep void in his life, a lack of ability to make ethical decisions. He seemed cut adrift, unable to trust what he heard and felt. Like his companions in government, he answered questions about Vietnam, Chile, or nuclear arms in the same way he answered those about his perceptions of his body: "Well, after all, how can one really say?" As such, he was the perfect civil servant. Since he could not trust the simplest of his perceptions, he could be counted on to obey government policies about more complicated matters, even if they seemed immoral.

Of course, Charles's mother's apparently trivial comment gained its significance for him because of widely known psychological factors. But to understand the full impact of her words on Charles's behavior, we need to look at the social milieu in which he was educated. Her scornful voice was augmented by a chorus whose fully unified hymn Charles heard throughout his life, wherever he might be. Other parts were sung by teachers, religious ministers, physicians, scientists, and politicians. The consistent theme that united their voices was that Charles did not possess sources of wisdom within himself, instead, there are external standards of good music, fashion, intelligent conversation, and what sorts of governments should be destroyed, and Charles should listen to officially designated experts who have the authority to set those standards.

Moreover, those repetitive voices were accompanied by images and body-shaping methods designed to evoke suspicions in Charles about the reliability of his own perceptions and feelings. He encountered these nonverbal messages in posture-training classes, in music and dance classes, when he learned to play polo, and as he learned how he should dress and move as an upper-class male WASP. That nonverbal invalidation of his sensual experience anchored the explicit teachings within his sinews, guts, and nerves.

Charles's case illustrates an important distinction between one's explicit ideas and one's behavior and emotional reactions. The age-old power of authoritarianism to shape people's lives despite repeated attempts at social change derives from both these sources, that is, convincingly argued theories about the unreliability of our perceptions and body-shaping methods designed to evoke conformity.

On an intellectual level, Charles rejected authoritarianism. He considered himself a child of the American Revolution, dedicated to populist ideals of personal freedom. But he acted and felt as if he were subject to people and forces beyond his control. Authoritarianism is more than an idea or a psychological force; it is woven into our bone marrow, causing us to behave like machines rather than self-regulating organisms.

Charles's case also illustrates what I mean by authoritarianism: a belief system holding that we must depend on the judgments of publicly designated experts for reliable decisions about personal and social life. Those experts, moreover, are thought to derive their authority not from the community they serve but from sources outside that community. Presidents and party chairmen derive their authority from classified knowledge kept from ordinary people for purposes of "national security." Scientists and physicians derive their claims from privileged access to experimental data and technical language. Popes, ayatollahs, and gurus are specially illuminated from above.

Pause

Recall a time when one course of action seemed clearly sensible to you but some authority belittled your judgment, making you doubt yourself.

Recall an incident when you had a sense or an intuition that you should do something out of step with popular opinion. Did you follow that intuition? If you didn't follow it, why not? What fears appeared?

Remember a significant period in your life when you had to make a major decision about your health, your profession, or your relationships. While trying to arrive at a final decision, did you take into account the feelings that were emerging in your body? Can you recall what those feelings were?

Think of strong feelings of anger, lust, and panic. Do these feelings make you consider your personal judgments unreliable?

TAMING OUR SENSES

In this section I am going to deal with one of the most subtle and pervasive assumptions about perception, which comes from a long history of literature that attempts to deal with the basic problem of why we go astray. People often make stupid judgments. Following their own insights, they commit atrocious crimes. Many turn to fanaticism. The greedy follow their selfish impulses to suppress the weak. Why is it, sages thoughout history in every culture have asked, that ordinary human beings fail to pursue their own best interests?

The curious thing about the classical answers to that question is their astounding agreement in singling out the human body and its perceptions as the culprit. That assumption is cross-cultural and transhistorical, cropping up in monasteries, research laboratories, and public schools. It supposes that what we sense — both in the outside world and within our own flesh — is unreliable, even danger-ous; it needs to be subjected to controls. Most ordinary human beings think that those controls exist outside them-selves.

My argument is that the classical assumption is too simplistic in identifying sensual knowledge as the primary source of human vagaries. Error, sin, and fanaticism can be found in all sources of knowledge: sacred books, mathemati-cal logic, technological instruments, popes, presidents and physicians. The fact that we have chosen the body and its sensations to blame has more to do with such unspoken issues as our fears of insecurity and death, men's fear of women, and the desires of those in power to maintain their authority — issues I will be examining throughout this book.

It is, of course, painfully obvious that we all go astray. Our personal judgments need some kind of checks to purge them of our biases, fears, and unenlightened self-interest. But how can that purge be accomplished? Are there more effective alternatives to the ways we've learned in our schools and churches?

In my own case, Roman Catholic theology had the earliest and most potent impact on my beliefs about the body. The kind of teaching I received as a young boy was a Latin example of a tradition that is found in cultures ranging from Ireland through Iran to southern India. One of its classic expressions is found in these words of Krishna to Arjuna in the *Bhagavad Gita:*

Yet by their violence, son of Kunti
The unharnessed senses
Can carry away the mind
Even of a wise man who tries.

Let him restrain them all
And sit steady, intent on me
For when he controls his senses
He can compose his mind.

When a man thinks of sensuous things
He becomes attached to them,
From that liking springs desire
And from desire comes anger.

From anger comes utter confusion
From confusion, wavering of memory,
From wavering of memory, loss of reason,
From loss of reason, he is lost.

But he, by ranging over objects of sense
With senses self-controlled, cut off
From desire and dislike, himself governed,
Attains a clear quietness of spirit. (2.61–64)[1]

With slight adjustments for cultural context, these words could just as well have been uttered by Socrates attacking the Athenian Sophists, Saint Paul preaching to his converts, or the Methodist minister down the street today. This theological division between senses and spirit, which unites otherwise disparate Baptists, Roman Catholics, Quakers, and Vedantists, creates a climate in which sensual knowledge is associated with sin and illusion. Even as a small child, I recall, I went to sleep at night afraid that I might sometime end up burning for all eternity in hell if I once let down my guard against following my own impulses. The world I saw and felt around me was far less real to me than the world of archangels and devils struggling within my soul to win its possession. Even the authority of the Bible was not sufficient to guarantee me a victory over the powers of darkness. Monsignor Renwald pointed out that our "passions" can cause us to twist the words of Scripture to fit anything we want, just like Luther and Calvin did. I had to rely on the directives of Bishop Armstrong and Pope Pius XII.

But Charles grew up in a world that scoffed at that metaphysical melodrama. His family was not religious. They espoused scientific humanism. His mother's scorn for his childhood response to Ravel formed the emotional basis for his adult scientific world view, which encouraged Charles to distrust his perceptions, not because they would tempt him into damnation but because they would lead him into foolishness and error.

Even though modern science derives from revolutionary attempts to get free of religious dogmatism, its pioneers failed to recognize the dualistic basis of that dogmatism. While Descartes and Galileo, for example, carved out a domain for independent rational inquiry, they continued to accept the prevailing division of the person into two distinct parts. Descartes called one part *res cogitans,* the "thinking thing," and the other *res extensa,* "the extended thing," the *corpus, le corps.* That division was born of his series of reflections on what he could be certain of, independent of outside authorities. He systematically eliminated all possible realities except for the "I" that was reflecting; only of that could he be certain:

> I am, I exist — that is certain; but for how long do I exist? For as long as I think; for it might perhaps happen, if I totally ceased thinking, that I would at the same time completely cease to be. I am now admitting nothing except what is necessarily true. I am therefore, to speak precisely, only a thinking being, that is to say, a mind, an understanding, or a reasoning being, which are terms whose meaning was previously unknown to me.[2]

It is difficult to discern what Descartes and Galileo and their successors bootlegged into the new science from the old theology, because dualism is such a universally accepted theory. Their assumption could be reworded something like this: As you sit in your chair reading this book, you do not have immediate contact with the cover of your book which touches the surface of your fingertips, with the dark lines on the white page through your eyes, or with the feeling of your back sinking into the chair. You "know" these realities only through remote neural connections integrated in the brain, which are accurately described only when they are expressed in quantified terms. All you directly know is the "I" to which you refer when you say, "I am sitting here reading this book."

Later thinkers denied even that tenuous knowledge. Descartes's successors, such as Locke, Hume, and Kant, pointed out the obvious conclusions of Descartes's logic. Since I do not have immediate contact with any of the realities of my ordinary life, I can be deluded about any of them — I could be dreaming, the images I see could be results of the residue of other images, my biases can be so strong as to distort what I feel, and so on. Only people who can use instruments and mathematics to describe those neural pathways and the atomic structures of their particles can claim any reliable knowledge about what I am seeing and feeling.

In this world view, the "thinking thing" is given over to the realm of literature and theology, since knowledge about it is thought to be based on faith or emotion. Rational truth belongs exclusively to empirical science, which examines a world reduced to those qualities that can be measured publicly and expressed mathematically: spatial dimensions and patterns of movement. My body, a *bottich* with feelings and primal wisdom, becomes a *corpus,* a collection of insensate particles moving in space. I can have reliable knowledge about it only through the methods of the new science.

Descartes's notion of the body has serious ethical implications. For Krishna, Plato, and Saint Paul, the human body had at least the significance of a dangerous animal; the ethical life required a courageous struggle to cope with its fierce passions. But the new science removed those passions — anger, love, lust — from the physical world, the body became a corpse, a collection of moving particles. Because it could reveal only mathematically expressed patterns, it could be a source of neither intellectual nor moral authority. "Value," incapable of being quantified, was relegated to the purely subjective realm of the private self, alienated from the human community. My clients Charles and Walter, for example, reflected this ideology in their impotence to make ethical decisions.

At first only a movement among European intellectuals, the scientific ideology slowly began to reform social institutions. By the end of the nineteenth century, the Cartesian gap between "I" and "my body" had become more than a bright idea. It was encrusted in the major institutions of society. Physicians, educators, and philosophers were teaching people that subjective experience of their bodies had no value compared to the objective knowledge possessed by a new class of experts. Moveover, the modes of work open to most people after the Industrial Revolution trained them to feel that their bodies belonged not to them but to their bosses.

Modern medicine, for example, originated in attempts to close the supposed gap between the perceiving self and the *res extensa* by using instruments, chemical analysis, and the dissection of corpses. People such as William Harvey and Louis Pasteur created a new system of medicine based on treating the human body like any other physical object, devoid of feelings, thoughts, and aspirations. "Health" came to be defined in terms of proper mechanical functioning of the various body parts.

Industrial capitalism was simultaneously becoming the dominant mode of production. The assembly line required bodies trained to behave with the predictability of the other mechanical parts of the process, so individual organic needs for rest and nourishment had to be forced into conformity with the standardized work week. If one wanted to retain one's job, one had to overcome sickness, disability, and extreme fatigue. Women had to have even more discipline, because of menstruation and pregnancy. A working force trained to have their bodies treated as property by management was gradually created, and freedom was left to the privacy of the soul.

Hegel and Marx called people's experiences of the Cartesian gap alienation: my body becomes a stranger to myself, something not truly mine. Such a discrediting of ordinary experience gained little popularity in the United States until the late nineteenth century, because it was associated in the popular mind with an un-American carryover from the European hierarchies of power and privilege. The ideology of alienation was brought to the New World primarily by a new kind of physician, one trained in biomedical research. How it arrived on these shores is an important story about how the body can be controlled in order to maintain the status quo.

I want to say at the outset that physicians play a particularly poignant role in the drama underlying this book. On one hand, as a group they have represented a consistent commitment to humanistic values. Doctors have been in the vanguard in advocating a need to respect people's resources for self-healing and in criticizing the industrial ideology for swallowing up doctors' basic commitments to healing. On the other hand, physicians are subjected to overwhelming pressures to convince ordinary people that they are incompetent to take charge of their lives. I do not know a single physician who does not feel torn between these two forces, one from the side of the people he or she is committed to serve and the other from those who would integrate medicine into the world of industry and politics. For example, in her review of a recent book on the manner in which the medical profession has been transformed in recent decades from a network of healing into an instrument of corporate profits, an internist chides the author for being too optimistic in stating that the complete corporate takeover of medicine may be forestalled, arguing that it has already taken place.[3]

According to our popular myths, health care in this

country underwent a dramatic improvement only after World War I, when the scientific medical schools such as Johns Hopkins began to flourish and the medical profession became highly organized. Before that, according to the myth, people had to rely on poorly trained local doctors and a variety of quacks. But the nineteenth century saw a rich variety of therapeutic methods reflecting the still vigorous populist spirit of the early Americans. Two feminist analysts of medical history write:

> The Popular Health Movement of the 1830's and 40's is usually dismissed in conventional medical histories as the high-tide of quackery and medical cultism. In reality it was the medical front of a general social upheaval stirred up by feminist and working class movements. Women were the backbone of the Popular Health Movement. "Ladies' Physiological Societies," the equivalent of our know-your-body courses, sprang up everywhere, bringing rapt audiences simple instruction in anatomy and personal hygiene. The emphasis was on preventive care, as opposed to the murderous "cures" practiced by the "regular" doctors. The Movement ran up the banner for frequent bathing (regarded as a vice by many "regular" doctors of the time), loose-fitting female clothing, whole grain cereals, temperance, and a host of other issues women could relate to. And, at about the time that Margaret Sanger's mother was a little girl, some elements of the Movement were already pushing birth control.[4]

Immigrants brought with them methods of healing that had been used in their cultures for generations and many people got help from the Native American healers. New Americans often developed their unique methods of healing. For example, Andrew Still created osteopathy, Daniel Palmer chiropractic. Samuel Thomson and Sylvester Graham (of Graham crackers) founded popular health movements based on diet, exercise, herbalism, cleanliness, and abstention from alcohol and coffee. Many of these people were vocal champions of organizing health care so that it reflected the American notions of participatory democracy, urging that people should have easy access to basic anatomical and physiological knowledge about their bodies and that healers should not insinuate European auras of privilege into their professions. A historian of medicine writes of that period: "Many Americans who already had a rationalist, activist orientation to disease refused to accept physicians as authoritative. They believed that common sense and native intelligence could deal as effectively with most problems of health and illness."[5]

During that era, healers placed emphasis on educating members of the family to take care of many diseases at home. Two popular books were called *Domestic Medicine*. The earlier, written by William Buchan, was described in its subtitle as "an attempt to render the Medical Art more generally useful, by showing people what is in their own power both with respect to the Prevention and Cure of Diseases."[6] The second, by John Gunn, maintained that Latin names for common medicine and diseases were "originally made use of to *astonish the people,* and aid the learned in deception and fraud. The more nearly we can place men on a level in point of *knowledge,* the happier we would become in society with each other, and the less danger there would be of *tyranny . . .*"[7]

Physicians were then more integrated into their local communities. There were proportionately more doctors than we have today, and they received modest wages like other artisans in their communities. There were several regional medical schools, more oriented to practice than to theoretical research, but many doctors learned the profession by becoming apprentices to older practitioners.

Within a few decades that diverse, community-centered system of health care was replaced by the chemical, technological, and financial empire we know today, with its vast network of hospitals. The widespread defense of the original American ideal of participatory democracy gave way to an Old World dependency on people of privilege. The transformation was so complete that in 1980, someone such as Dr.

Tom Ferguson, editor of a journal called *Medical Self-Care,* was considered a revolutionary for saying that most of us can take care of our health if we have some basic and easily accessible information about our bodies. The creators of industrial capitalism were among the principal agents of the dramatic change. The Rockefeller and Carnegie foundations created new medical schools where the emphasis was on biomedical research rather than clinical practice. The agents of these foundations, Abraham Flexner and Frederick Gates, lobbied to close the regional schools of practical medicine and to discredit alternative forms of healing as unscientific quackery. In the wake of Flexner's report for the Carnegie Foundation in 1910, six of America's eight black medical schools and the majority of the "irregular" schools that had been havens for female students were closed.[8]

A lucrative alliance began to be forged among the new medical elite, the corporate executives who financed their hospitals, insurance companies, and the industries needed to produce medical machinery and drugs.[9] This alliance was not so much a consciously engineered conspiracy as a result of what is now commonly called a paradigm shift, a change in the way vast numbers of people perceive reality. Several historical currents went into producing it. There were, for example, profound similarities between the forms of capitalism and scientific medicine. Both systems reflected the same philosophy of the universe. Capitalism, based on the economic theories of Adam Smith and the Social Darwinists, held that the success of production depended on rational analysis and control of all the discrete and impersonal components of the process: assembly parts, machines, rates of consumption — and workers' bodies, seen as quantifiable material units. The people qualified to hold power in this system were those who survived rugged histories of conflict to emerge at the top: Fords, Mellons, Rockefellers. They understood best how to manipulate the vast array of complex elements for the benefit of supposedly less intelligent humans.

Similarly, the medical elite saw the human body as a collection of discrete parts, the functioning of which was truly understood only by experts who knew biology, chemistry, and mathematics. They thought they knew how to repair the parts properly by surgery and intervene with drugs to destroy germs that were upsetting the system.[10] One historian writes that the creation of the new class of medical experts

> was no simple usurpation; the new authority of professionals reflected the instability of a new way of life and its challenge to traditional belief. The less one could believe "one's own eyes" — and the new world of science continually prompted that feeling — the more receptive one became to seeing the world through the eyes of those who claimed specialized, technical knowledge, validated by communities of their peers.[11]

The director of the Rockefeller Foundation, Frederick Gates, explained to Rockefeller that the body's industrial life exists in

> an infinite number of microscopic cells. Each one of these cells is a small chemical laboratory, into which its own appropriate raw material is constantly being introduced . . . The great organs of the body . . . are great local manufacturing centers, formed of groups of cells in infinite number, manufacturing the same sorts of products, just as industries of the same kind are often grouped in specific districts.[12]

Increasingly under attack by what was then a strong socialist movement in this country, the early American capitalists appreciated the respectability they found within the ideologies imported by European-educated scientists. Under the new paradigm the rapid accumulation of vast wealth by the new tycoons could be seen as something other than greed running unchecked; the new type of physicians and educators taught people to see it as part of a more effective organization of life itself. Hospitals and public schools now not only served as an effective network of social control, providing a steady supply of healthy and obedient workers for the factories, but they were also believed to improve the lot of the hundreds of thousands of immigrant workers in the major industrial cities of the New World. Thus, both industrial capitalists and the new class of physicians profited from alienation: the former from the alienation between the worker and his or her body, given over to the managers, and the latter from the alienation between the person and his or her body, under the doctor's authority.

Pause

Consider the authority your feelings have in your working life. How much weight do you give to the demands of your body for rest and movement?

Does the design of your workplace — chairs, desks, keyboards, work tables, video display terminals, etc. — support ease and flexibility in your back and neck?

Does your schedule allow proper digestion of food?

Is the lighting adequate to prevent eyestrain?

How much authority do you exercise over your health? Do you insist that your physician explain his or her diagnoses so that you understand them? Do you ask for alternatives? Do you feel that your own experience of your ailments is taken into account?

Sara is a forty-year-old mother of three children, a college graduate, well read and active in community affairs. Having experienced unusually intense menstrual cramps, she consulted her gynecologist, an older, gray-haired male and a Harvard Medical School graduate. He recommended exploratory abdominal surgery, in the course of which Sara's uterus could be removed, since, he argued, she no longer needed it. She asked him to explain what might be wrong with her and asked whether it would be good to get other opinions. The gynecologist was offended and curtly chided her for not trusting his opinion. She found her body slumping as she meekly apologized and agreed to have the surgery. It revealed nothing, leaving her scarred and without a uterus. Sara's far from atypical experience challenges those who might argue that medicine is not a dogmatic belief system like shamanism or witchcraft but is based on empirical research. The authority invested in physicians reveals how much more we are motivated by little-understood social forces than by evidence.

The improvements in our knowledge of the body from the nineteenth century to the present have occurred on many fronts. We now know more about diet and exercise than Samuel Thomson and Sylvester Graham did. Osteopathy and chiropractic have become more sophisticated, their schools more rigorous. Various forms of manipulative therapy have been imported from Europe, some have been developed in the United States. Asian medicine has made an enormous impact on health care since World War II. Anthropologists have made us realize that the popular discounting of folk medicine as "superstitious" is based on myth rather than evidence, and many physicians now regard native healers as their peers.

Psychoanalysis has called into serious question concepts of disease based solely on the old mechanical model of the body. Large numbers of women are once again undertaking training as midwives. Caring for the sick in their own homes is now seen as more effective than transporting them to hospitals, except in extreme cases. Dr. Eva Salber, a professor of community medicine, writes: "The great majority of illnesses are never seen by a doctor. The *real* primary care is provided by one's family, close neighbors, and friends. In every community there are people others turn to for advice, counsel, and support. I call these people health facilitators. One of the most important things doctors and other health professionals can do is to *find* these facilitators and offer them recognition, information, and support."[13]

Many improvements in bodily life are due to modifications in nutrition, sanitation, and working conditions, along with reductions in family size. "In the great majority of cases the toll of the major killing diseases of the nineteenth century declined dramatically before the discovery of medical cures and even immunization."[14] René Dubos writes: "The tide of infectious and nutritional diseases was rapidly receding when the laboratory scientists moved into action at the end of the past century. In reality, the monstrous specter of infection had become but an enfeebled shadow of its former self by the time that serums, vaccines and drugs became available to combat microbes."[15]

Of course, along with all these other advances, biomedical research also had its stunning discoveries, such as insulin, antibiotics, polio vaccine, and L-dopamine. Surgeons developed extraordinary methods for repairing hearts, eyes, and brains, while orthopedists devised ways of making old age more comfortable through sophisticated prosthetic devices such as knee and hip joints. But in context, the proven successes of the new medicine are far more modest than one might expect from the authority it enjoys in our culture. In proportion to the money spent and the social status of physicians, the results of scientific medicine in improving the health of the ordinary person are meager.

The transition from the nineteenth century to the twentieth was not simply one of improving our bodily lives; it involved consolidating the centralized power of industry and weakening democratic ideals that might threaten that power. Biomedical science, based on denigrating the value of people's own experience and emphasizing individualistic health factors, neatly fit into a growing network of social controls. It benefits industrialists to have their workers believe that the principal sources of bodily discomfort are to be found within the individual: laziness, sickness caused by germs, scoliosis, and neurosis. To emphasize the social factors in bodily well-being might erode the base that supports the economy. Corporate executives have consistently lobbied against legislation to regulate the safety of workers, protect the environment from industrial pollution, support research in nutrition, and promote safeguards against consumer deception.

The present system of health care, which is capital-intensive, significantly contributes to industrial profits by emphasizing a network of technologically and drug-oriented hospitals. Cutting up bodies and altering their chemistry is a profitable industry. An adult like Sara, successfully educated in this system, has learned the impropriety of questioning the doctor's prescriptions of major surgery or drugs. "The obedience of a patient to the prescriptions of his doctor should be prompt and implicit," reads the first code of ethics drafted by the American Medical Association. The patient "should never permit his own crude opinions as to their fitness to influence his attention to them."[16]

During the past fifty years, the marriage between corporate power and scientific medicine has encouraged our passive relationship to our bodily existence. Parents, teachers, coaches, and nurses support the biomedical ideology, training us that our self-perception counts for little. The doctor knows what is truly good for my body.

CASTING SENSUAL MOLDS

The mind-body dualism that supports authoritarianism is more than a verbal, explicitly taught theory. In fact, such doctrines would probably have little effect if they were not supported by a wide variety of body-shaping techniques that train people instinctively to look outside themselves for direction.

I've described how both Walter and Charles shrank in the presence of authorities. Wilhelm Reich called those postural collapses "the physiological anchor of the social incapacity for freedom."[17] Little matter that these men were physically strong, intellectually alert, and dedicated to ideals of personal freedom. When it came to taking hold of their lives, they were impotent. Their loss of stature is merely one instance of the embodiment of authority in our society: the visibly deferential ways in which adults shrink and smile in the presence of those they regard as superior, the ways in which children and teenagers learn to posture around adults, the nonverbal games men and women typically play to support sexism. These forms of behavior are not haphazard. They result from a well-orchestrated system of methods for shaping bodies in our culture, which we will examine in the next three chapters. At this point I will give a general survey of the areas at which you might look to find the implicit teachings that permit the authoritarian anchor to enter our flesh.

From infancy through old age we are taught to conform our bodies to external shapes. We learn to perform physical activities in specifically prescribed ways. We are rewarded for keeping quiet and controlling our bodily impulses. The implied meaning of these recurrent nonverbal messages is consistent with the explicit teachings: our bodies, with their feelings, impulses, and perceptions, are not to be trusted, and must be subjected to external controls to keep them from leading us astray. They must be trained to support the status quo.

Status comes from the Latin word that means "standing." *Status quo* refers to the stance one took previously, so maintaining the status quo means to keep standing in the same place even though everything else changes. Training people to maintain it requires educating them to stand still.

I'm speaking here of a more precise phenomenon than a "habit." Status carries both a physical and a social connotation. One strives for status within different hierarchies, and once one achieves that place one must adopt the kinds of movements and postures appropriate to that status. Thus stance is not an ephemeral thing; it leads to shaping muscles and bones so that a person will automatically adopt predictable forms.

Pause

When you are working, do you cling to certain postures even though you customarily experience discomfort?

In your ordinary activities, do you allow yourself to move and stretch from time to time, or do you remain relatively

fixed in certain positions? What kinds of inner voices keep
you from being more mobile?

Think of different social situations in which you find
yourself: at work, home, your parents' home, church, business meeting, parties with people you don't know well, and
so forth. Do you find yourself using postures in one place
that you change in others? Do you notice any similarities
between changes of clothes for these different scenes and
changes in how you use your body?

The problem with the status quo is that many of our personal and social conflicts come from confronting new situations with old stances. Holding on to old postures may have
relatively trivial results, as when I continue to lift my eyes
from the golf ball even though thousands of flubs have
taught me the inadequacy of using my body that way.
Keeping to the status quo becomes more serious if I
repeatedly, almost mechanically lose my temper with my
stepdaughter when she fails to clean up the dishes, since it
obscures my love for her and makes her recoil from me.

At the social level, we keep to old modes of production
and defense when they are no longer useful. Movements that
once felt right now seem out of joint. We feel like machines
programmed to act in set ways even though our ideas have
changed. Reich wrote of the social education of the body:

> Bringing people up to assume a rigid, unnatural
> attitude is one of the most essential means used by a
> dictatorial social system to produce will-less, automa-
> tically functioning organisms. This kind of upbringing
> is not confined to individuals; it is a problem which per-
> tains to the core of the structure and formation of
> modern man's character. It affects larger cultural cir-
> cles, and destroys the joy of life and capacity for happi-
> ness in millions upon millions of men and women. Thus,
> we see a single thread stretching from the childhood
> practice of holding the breath in order not to have to
> masturbate, to the muscular blocks of our patients, to

the stiff posturing of militarists, and to the destructive artificial techniques of self-control of entire cultural circles.[18]

Reich was writing of Hitler's Germany, where there was a clear connection between the shaping of individual bodies and social ideology. Today it's easy to see such a relationship between the body and the body politic in totalitarian societies such as China and Russia, which dictate uniformity in body training, dress, diet, living conditions, and political ideology. It is also clear in more traditional homogeneous societies. In Africa, for example, the Ibos begin molding a child's head at birth "because round heads are usually preferred, pointed ones being ill-adapted for carrying loads in later life."[19] Peasants in Tashkent, on the other hand, prefer elongated heads.[20]

One might think that the cultural diversity in the United States precludes any possibility of establishing a relationship between individual bodies and social forms. Our shapes manifest all the styles of training and artistry found throughout the world, and our bodily education is heterogeneous, involving different paths of exercise, diet, and work. There are at least as many kinds of bodies as there are religions and political opinions. But despite the illusory maze of individual and social shapes, we are not all that different from the Germans of the Third Reich, the Russians, the Chinese, the Ibos, and the Tashkent peasants. Our bodies too are molded to bear the social load and to manifest chauvinist ideals. But in our more complex society, the nature of the molding process is obscured by popular myths of American pluralism. Even in our supposedly permissive, democratic society, bodies supple at birth are slowly trained to become rigid and stereotyped in their responses to life. We have a widespread network of methods designed to shape instinctual impulses for the sake of preserving the status quo.

The prototype is military training. Cadets are taught definite forms of holding themselves, precise styles of walking, sitting, and standing. Spontaneous behavior is implicitly sanctioned only within specified limits of time and space: in dark rooms and anonymous bars. Deferential physical behavior toward authorities is demanded. Dress codes, the design of furniture and classrooms, the styles of marching music, forms of exercise, and competitive sports contribute to the desired body image. The result is a body in which all organic impulses are rationally subjected to the demands of one's superiors. Similar to all the other bodies in the academy or boot camp, the person is given the nonverbal message that individual differences are not of value. What is important is the group, which insures protection from the enemy without and rebellious instincts within.

Penal institutions — both prisons and the older types of mental institutions — use similar techniques to achieve standardized behavior in people judged to be deviants from social norms. The role of these institutions has been to shape individuals to traditional norms by using drugs, solitary confinement, surgery, and torture. In his history of mental institutions, Michel Foucault writes that

these great internment houses were created with the intention of receiving not simply the mad, but a whole series of individuals who were highly different from one another, at least according to our criteria of perception — the poor and disabled, the elderly poor, beggars, the work-shy, those with venereal diseases, libertines of all kinds, people whose families or the royal power wished to spare public punishment, spendthrift fathers, defrocked priests; in short, all those who, in relation to the order of reason, morality, and society, showed signs of "derangement"... The common category that grouped together all those interned in these institutions was their inability to participate in the production, circulation, or accumulation of wealth... Internment, therefore, was linked in its origin and its fundamental meaning, with this restructuring of social space.[21]

The extremes of standardizing bodily behavior and numbing instinctual urges that have characterized military schools, prisons, and mental hospitals merely put into relief the less obvious forms of the same processes that occur in seemingly more benign institutions. For instance, schools are primarily designed to train docile citizens and workers. This end is achieved by definite forms of bodily behavior. Small, highly mobile children are made to sit in rigid desks for hours without significant movement. They are to speak or move from their place only by going through a prescribed ritual of raising their arms and being recognized by the teacher.

Think of the number of hours in your life you have spent sitting quietly in desks or chairs in rows facing straight ahead in the direction of some expert or another. You might be able to recall memories of how your body feels in such instances: its tensions, restlessness, sluggishness. Think of how little perception you have of other students, workers, and so on in such configurations.

Bodily patterns of fatigue, hunger, and excitement are brought into alignment with the externally determined rhythms of the school day. Art and sports are carefully regulated, and students are told how they should draw, mold clay, and throw a ball. Idiosyncratic behavior is generally punished, either physically, in the case of students who are too loud or restless, or through poor grades, in the case of those who don't express themselves "correctly." Many young children are given drugs to suppress a hyperactivity that might often be the result of the rigidity of the classroom rather than of personal organic disorders.

Industry is a principal beneficiary of these corporeal disciplines. Schools train people in the bodily patterns that most jobs require. The organic rhythms of the body are geared to meet the needs of a standardized working day, beginning and ending at certain hours, with carefully specified breaks for food, toilet, and rest. For both factory and office workers, body movement takes place within carefully defined limits set by industrial engineers to maximize efficiency. A typical handbook argues: "There is but one way in which the office manager can control scientifically; that is by standardization

. . . The office manager should, therefore, continually direct his efforts to having each operation. . . always done in exact accordance with the manner he has prescribed."[22]

The stress provoked by this environment has not gone unnoticed. Most of our bodies, with their odd shapes and sizes, balk at being packed or stretched to fit society's procrustean bed. Some of us react to that stress individually: we get headaches, backaches, or heart disease. Others react collectively, forming unions or cooperatives.

But those in power are resourceful in maintaining power. Corporations offer tranquilizers through their nurses, sponsor psychological seminars designed to promote cooperation between workers and management, and are increasingly making use of newly developed stress-reduction techniques.[23] Health-insurance companies are designing health-engineering programs, which they intend to fund in major universities, with the goal of producing experts whose job would be to train workers in how to relieve their stress. This movement is a contemporary version of Rockefeller's and Carnegie's funding of medical schools on the European model to augment their control over production.

A symbol of the new health ideology is a stress-reduction device called the "Environ Personal Retreat." It is a totally enclosed cylinder, featuring a cushioned armchair, rose-scented and ionized air, soft colors dancing on the ceiling, and various taped messages of relaxation and prosperity consciousness. Lulled by such sophisticated narcotics, we lack the vitality that would enable us to exercise power and create a world more suited to our yearning flesh.

A PERVERSION OF VALUES

The symbiosis of ancient theories invalidating sensual authority with nonverbal body-shaping methods aimed at shaping individuals for social goals has produced societies motivated by curious hierarchies of values. Think of us as a community of biological organisms supported within a natural environment. Think of what it takes to sustain and enrich this community at its most fundamental level: food, shelter, communal organization of work, care of the sick. Then think of the most abstract human goals: an Islamic society, communism, capitalism, Christianity. If you examine our social goals you will find that we are collectively supporting the most obscure goals, whose "truth" is difficult to verify, and sacrificing the pursuit of those that are the most obvious and tangible.

A microcosm of planetary insanity, Ireland is a poignant example of how this perversion of values operates. The common plight of the Irish people is economic: they don't have enough food, land, or control of their own industries. Centuries of British ownership of land and industry have sapped the country of its resources to support itself. Instead of banding together to reconstruct a society that serves their needs, the people pursue old ideological battles. A Catholicism imported from third-century Rome does battle against a Presbyterianism imported from seventeenth-century Scotland. The participants are so mesmerized by these mystical ideologies that they cannot confront their obvious problems. Their society is sick in the sense defined by Herbert Marcuse: "Its basic institutions and relations, its structure, are such that they do not permit the use of the available material and intellectual resources for the optimal development and satisfaction of individual needs."[24]

It is easy enough to notice the abstract nature of the values of those we label extremists: Nazis who betrayed their families and neighbors, Iranian revolutionaries who turn their children in to the Islamic courts, white racists in the United States who harass blacks, Jews, and Latinos. It is more difficult to recognize our own perversions, to look at the implications of the fact that with our votes and taxes we support governments that have allocated so much wealth to defense that we cannot meet the fundamental biological needs of the human organism. It is not farfetched to compare those we label extremists with the leaders of the United States and Russia, who talk of risking the existence of human life itself to preserve their beliefs.

One can argue that all these groups, extremist as well as "civilized," are not really motivated by abstract ideologies, but by survival. But the paradox remains: Our notions of survival have become so distorted by our vague fears that we think in terms not of negotiated allocations of food and land but of the extermination of millions of people. In our need for survival, we have given up our collective genius, renouncing the sensually obvious for the abstractly obscure. I think of this book as a handbook of exercises for recovering that forgotten genius. It is designed not to convince you of my theoretical point of view but to stimulate you to discover your own unique sources of wisdom and to gain some power over the barriers that keep you from acknowledging that wisdom.

Genius comes from the Latin word that means one's unique spirit, my unique perspective on the world as distinct from yours, what I alone can contribute to life. It emerges in those moments when we come to our senses, expressing ourselves spontaneously, it is a djin released from Aladdin's lamp to perform all sorts of wonders. There is also the "local genius," the unique characteristics of life in Vermont villages or Lower East Side Manhattan or towns in the Texas Panhandle. And there is the genius of a people: Bretons, Haitians, Balinese. Genius bubbles up from the brewing vat, the *bottich,* the body. It refers to the finest distillates of an individual's or a culture's sensual life.

It is no accident that in authoritarian societies, "genius" is applied only to certain individuals who, according to the norms of that society, are thought to be more gifted than others — people who understand particle physics, who have access to secret government documents, or who, like Charles's mother, know that Ravel is trashy. Authoritarian structures are supported by populations of suppressed geniuses, people who have been educated from infancy to distrust their sensibility. Such people easily attribute their failure to comprehend public policies that seemingly contradict fundamental human values to their own lack of expert knowledge of the issues. Or, trained in passivity, they despair of affecting such policies.

Throughout this book I am inviting you to become more aware of the enormous treasury of immediate, authoritative data you possess about your body and the sensual world. You can easily become intimately aware of your stress patterns, the distinctive ways you move your hips, the kinds of food and exercise that benefit or harm you, the emotions associated with different uses of your body. You have perceived as much or more than I have about the bodies of others. You have watched people on the streets, noticing their peculiar styles of walking and gesturing. Perhaps you have noticed the differences among various ethnic groups, especially if you live in a large, cosmopolitan city or have traveled widely. If you're a parent, you have learned about body development by watching your children grow. With lovers, dance partners, and athletic companions, you have noticed how movement patterns in other people affect your sense of yourself.

But if you are like Charles, Sara, or myself, you may not have given much, if any, weight to these data, thinking that you know very little compared to coaches, gynecologists, and dance instructors. This book is an invitation to recover connections with your sensual knowledge so that with our shared genius we can find our way back across the ever-widening gulf between our social policies and their biological substrate.

CHAPTER THREE

MY BODY / MY POINT OF VIEW

. . . in this intricate, dense, moist web of cells we carry
around with us (and not in any airy thing attached to it)
lies the substance of all the love and hate, joy and grief,
hardheaded analysis and excited imagination
we experience during our sojourn on this planet.
And it is only because certain cells make signals,
chemical and electrical, to communicate with each other,
that we are able to think and feel at all.[1]

This book is necessarily "complicated" in the radical sense of having many folds and wrinkles. Its case histories, autobiographical sketches, anatomical analysis, and psychological and social theories are meant to reflect the complexity of our immediate experience. At one moment a person may be walking down the street, worried about getting to his office on time. He may also be noticing the discomfort of a mild hangover. He thinks about a story in the morning paper about another massacre in El Salvador and feels disturbed. He notices an aftertaste of coffee in his mouth. He remembers an interesting theory he read yesterday in a book about the brain. He suddenly plunges into depression as he adverts to the fact that he is in the midst of the breakup of a twenty-year-old marriage. We are trained to think of these prolific ramifications of our sensual experience as atomistic pieces. The theme of this book is the process of reconnecting these fragments by showing how they are rooted in our bodies.

Our perverted social values derive in part from our training, which encourages us to feel that our points of view, our ideologies, are one thing and our bodies are another. Accepting this metaphysical fracture as natural, we fail to notice that our points of view are constantly being shaped to support social trends that are not necessarily in our best interest.

One of the problems you may have in following the argument in this book is that we are so immersed in a dualistic understanding of the world that we instinctively identify "ideas," "attitudes," "inclinations," "tendencies," and "biases" as mental or psychological realities, disconnected from muscles and nerves.

But *attitude,* for example, comes from the Latin *aptitudo,* which means the way one "fits in." It evolved into the word that meant the angle at which one is inclined from the vertical — the way one is tilted.

Pause

If you observe yourself in a full-length mirror, drawing an imaginary line through the middle of your brow to the tip of your nose, the center of your chin, the base of your neck where your collarbones join, through your navel and the center of your genitals, and down between your knees and ankles, you will notice that the line wanders, now to the left, now to the right. If you have a friend, your husband, wife, or lover stand next to you and draw a similar line, that line will take at least a subtly different course. These differences reflect unique torsions in the connective tissues of your body, its muscles, tendons, ligaments, fascial sheaths, and bones.

You may also notice characteristic asymmetries: your right shoulder may be slightly larger and higher than and forward of your left. If you look directly down, your left thigh may appear to be a little behind your right. Your left foot may be a half-size larger than your right. Your partner's contours will be different.

Try to find pictures of yourself as an infant standing naked or in something revealing like a bathing suit. You may notice some of these same patterns even at that early age.

Those two different imaginary lines and contour patterns will be reflected in the different ways in which you and your partner see and touch. They will emerge in the different ways you saunter down the street, dance, play sports, and make love. You each "fit into" the same chair and conform to social norms in characteristically different ways.

A *tendency* originally meant a "stretching out toward." A *propensity* or *bias* represented "weighted pulls" toward this rather than that. A *bent*, of course, resulted from a bend.

You may sometimes notice that you muscularly shrink away from certain people, or feel a sinking feeling in the pit of your stomach when you contemplate taking a difficult stand, or feel an overwhelming fatigue after a stressful meeting, or feel high-strung anticipating an uncomfortable decision.

Only in recent decades have we begun to understand how those different tilts, pulls, and sinking feelings are reflected in our more abstract religious, philosophical, and political differences. The cases of the following women illustrate the subtle connections among muscles, perceptions, and more general attitudes toward life.

TWO POINTS OF VIEW

Sister Ruth is a fifty-year-old Franciscan nun, a hospital administrator from the Midwest. Imelda is a forty-six-year-old social worker and political activist from California who is divorced and has one son, a blind twenty-six-year old. In a recent training seminar for psychotherapists, these women were asked to draw their bodies as they experienced them. Sister Ruth drew this:

Figure 1

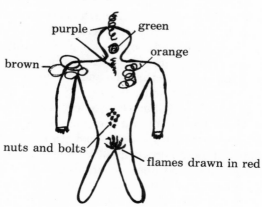

Sister Ruth is of German descent, her four grandparents having emigrated to the United States near the turn of the century. All of her family, including her youngest nieces and nephews, are devout Roman Catholics. She described the strong black outlines of her picture (see Figure 1) as expressing her feelings of clear demarcations between herself and others, a highly defined sense of where her body ends and the rest of the world begins. She felt that this sense of herself came from her experience of her family, where everyone's role was clearly defined, her father being the rational head and her mother the warm source of affection. The behavior expected of each was clearly spelled out by the laws of the Church.

Sister Ruth's gait is heavy and defined; she holds her arms close to her sides, sometimes with her fists clenched. She has worn foundation garments most of her adult life. She expresses herself clearly and directly.

During her late twenties, Sister Ruth gained nearly one hundred pounds and coincidentally developed severe back pain. In 1963, she had the first of four spinal surgeries, none of which was successful in eliminating the pain, which eventually spread into her legs and feet. She became addicted to pain-killing drugs to such an extent that she had to undergo withdrawal therapy in a clinic. In 1975 she had a hysterectomy, and in the spring of 1980 she had an operation in which her stomach was stapled, after which she lost seventy-one pounds. The nuts and bolts in her drawings refer, she explains, to places where she feels rigidity in her body. She was surprised to realize that she had drawn them in the same areas where she had had surgery.

Sister Ruth's body has always been a conscious presence because of the back pain and the excessive weight. In addition, her vow of celibacy required constant attention to suppress what she says were tormentingly strong sexual desires. The boldly drawn red lines around her vagina express the power of those desires and the energy required to keep them from spreading outward into other parts of her body and eventually into her behavior. She says she was always afraid of those desires; she knew that permitting them would not only lead her away from the convent but would also entail her ostracism by her family, particularly her father.

Here is what Imelda drew:

Figure 2

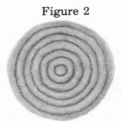

The darker colors are orange fading into light yellow.

Imelda's cultural background covers several continents. Her maternal grandfather was a Cherokee, both her maternal grandmother and her paternal grandfather were a mixture of black African and German, and her paternal grandmother was Irish.

In contrast to the defined hierarchy in Sister Ruth's family, the structure of Imelda's family was like that of a team. She says that its tightly bound unity, which allowed no sense of separate roles, made it difficult for her to gain any sense of personal identity. Though her family regularly attended a Baptist church in the Texas town where she grew up, she says religion was more like a social activity; her parents told her that it made no difference what church she went to, but it was best to stay with the family. She had childhood friends from many religious backgrounds, including Buddhists.

Imelda describes her drawing as an expression of her feeling that everything is related to everything else. The orange lines fading into yellow and blurring into the white of the page show her sense of continuity between her body, the world, and the "spiritual." Physical shapes are not "real" for her, in the sense of being clearly defined and separate objects.

Imelda is soft and curvaceous and usually dresses in loose-fitting clothes that give the impression of gliding when she moves. She speaks in roundabout ways; people complain that they have difficulty getting a handle on what she means.

In 1963, suffering from an acute asthma attack, Imelda was rushed to the hospital, where she was mistakenly given a drug to which she was allergic. To all appearances she choked to death, showing the clinical signs of death for nearly twenty minutes. During that time she experienced herself floating above her body, observing it peacefully. She was eventually drawn into a long tunnel where she began to feel intense bliss and to see orange and yellow colors similar to the ones she drew. At one moment she felt she was given the opportunity to remain in that state of bliss or to return to the world to help other people. She chose to return. She says that she has retained the sense of unity with the cosmos she felt during those moments and expressed in her drawing.

In 1973 Imelda contracted uterine cancer, in the course of what she describes as a guilt-ridden affair with a married man. Rejecting medical advice to have a hysterectomy, she says she cured herself by leaving the man and getting rid of the "negative thoughts" that supported the cancer.

Imelda's body has not imposed itself on her awareness

with the same vigor as Sister Ruth's has. She feels she is passing through bodily life, not anchored in it. The colored lines fading into each other mean to her that there is no clear demarcation between what she experiences as "bodily" and what she calls "psychic" or "spiritual." For her, sex and religion can take many forms, no one of them being particularly more valid than any other as long as love is present.

Sister Ruth and Imelda express their ultimate life goals similarly: they wish to spread love in the world, assisting people to realize their inner essence. But what that means for each of them is very different. Imelda perceives people in unusual ways that reflect her self-perception. She has visions about them, grasps abstract patterns, sees auras, has intuitions about future events. She is training as a psychologist to learn how to integrate these perceptions in a way that will be useful to others. She intends to continue her work as a political activist, working with minority groups and the peace movement.

In contrast, Sister Ruth intends to direct her work primarily at Catholics, teaching them how to improve their spiritual lives through exercise, sports, manipulative therapies, and awareness techniques. She will work through existing Church structures: her religious order and its retreat houses.

Imelda's and Sister Ruth's drawings, body structures, styles of movement, religious beliefs, family histories, and life goals are all of a pattern. That pattern, however, is not a logical series in which one level can be deduced from another, so that one can say, for example, that Sister Ruth's scars caused her to draw the nuts and bolts where she did, or that Imelda's nonauthoritarian family structure blurred her perception of bodily boundaries. The pattern that connects the various levels is aesthetic.

Aesthetic, from the Greek, originally meant "perceived" "sensual." An aesthetic pattern is one we grasp with our senses. It is like a patchwork quilt or a Tibetan Thangka in which several pieces form one visual image.

Part of that pattern is what both women mean by "body." For most of her life, Sister Ruth thought of her body as the classical *corpus*, a clearly delineated, tangible object apart from her immortal soul, the part of her in contact with God. Her "felt" presence of God, however, always made her dubious of that division, and her recent experiences of her body's engagement in prayer have caused her to rethink the notion. When Imelda says she doesn't attach much importance to the body, she also means the body as *corpus*, a dense thing separate from other objects "out there" in the world. But her experiences are of pulsations, vibrations, colors, extraordinary perceptions, felt connections — all of which are open to neuromuscular interpretations. They are compatible with the ancient notion of the body as *bottich*, the brewing vat of spirits.

These women's notions of "authority" are part of the pattern. Sister Ruth is a witness to the profound effects of masculine authoritarianism on the body. She has been shaped by male "experts." She was taught as a child to regard her father as her fundamental authority, her most immediate contact with God the Father. Male confessors instructed her in how to keep her bodily impulses under control; male physicians carved her muscles and stapled her guts; a male pope and male bishops regulate every detail of her life, down to the kinds of clothes she is allowed to wear. Imelda, on the other hand, lives on the fringes of authoritarian structures. She relies on her own perceptions. Her own father exercised little authority over her when she was a child, and she has little to do with physicians, religious ministers, or husbands. It's not that she is rebellious; she just doesn't seem to care much about what the "experts" say.

THE BODY: SUBJECTIVE AND OBJECTIVE

Like Sister Ruth and Imelda, each of us views the world from a unique vantage point, standing at peculiar angles on feet of varying shapes. Our different breathing rates, walking styles, and muscular shapes present different pictures to those who watch us. Friends will recognize us coming a block or two away, before they can even see our details.

Our abstract belief systems form a piece with our more perceptible qualities. Your reactions to this book, for example, are not simply the result of the transmission of impressions of lines from your eyes to your brain, where certain ideas are triggered. Your whole body transmits data and organizes meaning. You may be remembering events from your past which evoke a variety of impulses and emotions. Your comprehension of these lines will be a function of how rested, hungry, and comfortable you are. As you read, you may shift your position or walk around to relieve fatigue. Some passages may cause you to tighten your belly in anger; others may pique your interest so much that you forget all the background noises and bodily discomforts that were making concentration difficult a few moments ago. Some passages may evoke sexual feelings, others fear. Poor lighting may cause stress in your eyes which will radiate into your head, neck, and shoulders. The subsequent discomfort will make it difficult for you to pay attention to the reading.

I have designed this book, with its frequent autobiographical references, to give you a feeling for how these "ideas" have bubbled up from my unique bodily processes. As you get a more precise sense of how my viewpoints are fermentations of my rigidities, shapes, and ranges of movement, you will be freer to judge the value of my ideas in relation to your own propensities. Every book represents a distillation from the author's *bottich;* every argument is *ad hominem.* But many books don the garb of "objective" language, lulling us further into the sleep of dependency on "those who know."

Meaning — of quantum mechanics, a painting, "life" — is one dimension of our bodily interactions with each other and with a sensible world. Some of those interactions are direct, as when we touch a lover or taste a cup of coffee. Others are indirect. You and I communicate via printing press and paper. We learn of the stars through telescopes, of cells with microscopes. Within the network of bodily encounters I become aware of certain attractions and I form the projects that will shape my life, pursuing certain people and goals. Repelled by certain smells, sounds, or inner feelings, I form my prejudices. My world view is first of all literally a viewpoint, my particular angle on life, derived from where and how I stand, reflecting my peculiar leanings and those biases tipping my scale toward one kind of behavior rather than another.

But in recent centuries people have not been accustomed to think of their bodies as active sources of meaning. The prevailing conception of how the human body functions in the formation of ideas and values could be compared to that of a camera in the photographic process. The Cartesian body, the "extended thing," was thought to have a definite structure described solely by the laws of physics. People pictured stimuli passively received from external objects as mechanistically processed through the nervous system to form an image in the brain. Then, by methods varying according to one's epistemological bias, the image was used by the "mind" to form ideas.

The turn of the century witnessed a profound revision of Descartes's mechanical notion of "body," which occurred among a group of European neuropsychiatrists, Gestalt psychologists, psychoanalysts, and phenomenologists.

Sir Henry Head in England noticed in his work with wounded soldiers certain phenomena that could not be explained by the traditional model. Although one patient, for example, had a leg amputated, he continued to experience accustomed movements in the location of that foot and leg until the time when he had a stroke. Without the brain lesion accompanying the stroke, the person continued to have a stable sense of his body, even though it had been radically altered. Head argued on the basis of several such experiences that every individual has a relatively continuous image of the position of his or her body and the relationships among its various parts. That image, embedded in the sensory cortex, may persist even when the "objective" body has been changed by something as drastic as amputation. He called

this image "the postural schema, of the body," the underlying standard to which an individual compares any new
movement and sensation in making the simplest judgments
about his or her own position and the location of objects.[2]
Sigmund Freud's early work consisted of research into
the systematic inadequacies of "objective" or neurological
explanations to account for certain kinds of symptoms. In
1886 he wrote a ground-breaking study of a hysterical male
which overturned the traditional dogma linking that disorder
to women. In that study he showed how hysteria looks dis-
cernibly different from organic disturbances of the nervous
system. Organic disturbances obey neuroanatomy; hysteria
"takes the organs in the ordinary, popular sense of the names
they bear: the leg is the leg as far up as its insertion into the
hip, the arm is the upper limb as it is visible under the
clothing." Hysteria behaves as though anatomy did not exist,
or as though it had no knowledge of it.[3] Freud's hysterical
man, for example, dragged his whole leg like an inert mass,
while the person afflicted with the pattern of brain damage
called hemoplegia performs slight movements of rotation in
the hip, movements totally explained by lesions in certain
pathways rather than in others.[4] Unlike organic paralyses,
whose causes can be predictably located on the neuroana-
tomical map, hysteria indicates some "injury" to the person's
conception of his or her body or body image. In other words,
for the hysteric, "the hand," paralyzed and unfeeling, is not
a system that is innervated by definite pathways and fed by
a definite system of arteries; it is the object that his mother
slapped when she caught him masturbating.

In the early 1900s, the German Paul Schilder began his
life's work of research into body image, culminating in his
The Image and Appearance of the Human Body. He discovered that any person's perception of his or her body contained
infinitely more than could be described by any so-called
objective account. It summarized a person's psychological
history and expressed his or her involvement in a social
world, influenced by such factors as dance, fashion, gymnastics, and styles of expressive movement. What any individ-

ual means by "my body" is not limited by the borderline of the flesh and its clothes; it

> can shrink or expand; it can give parts to the outside world and can take other parts into itself. When we take a stick in our hands and touch an object with the end of it, we feel a sensation at the end of the stick. The stick has, in fact, become a part of the body-image. In order to get the full sensation at the end of the stick, the stick has to be in a more or less rigid connection with the body. It then becomes a part of the bony system of the body . . . [5]

Schilder tells of an automobile accident in which he had a severe injury.

> In the early days after the accident every approaching car seemed to involve a particular danger element which encroached into the sphere of the body, even when it was a considerable distance away. In other words, around the body there was a zone closely interrelated with the body-image which was in some way an extension of the body. Later on this general zone diminished in size until finally there remained only a zone around the painful hand.[6]

As an example of how the space of "the body" cannot be explained simply by physics, he describes the case of a woman whose nine-year-old daughter had been run over by a truck a year before the woman's first admission to Schilder's mental ward. As he reports it, the woman described her experience as if her body were spread out over the surrounding environment: the street-cleaner five stories below her window sweeps over her genitals; another man "breathes her breath;" a doctor "walked on top of me; he was stepping on my body (he was not near me, but it hurt me)"; when Schilder talks "unnaturally," he hurts her "here and there"; he touches her with his cough, moves her shoulders by moving his own, etc.[7] "In every action," he writes, we are not only acting as personalities but we are also acting with our bodies. We live constantly with the knowledge of our body. The body-image is one of the basic experiences in everybody's life...Whatever we may do we want to change the spatial relation of the body itself. When we see something, muscular actions start immediately, and at once bring about a change in the perception of our body. Every striving and desire changes the substance of the body, its gravity and its mass.[8]

Schilder found that many people have a fluctuating sense of their body size: sometimes it seems massive, at others infinitesimal. One part of the body may be felt as distorted in relation to others, as when Sister Ruth drew her shoulders as much broader than her hips, when her hips actually are broader than her shoulders. Schilder discovered that some people have a sense of lines of force connecting parts of their bodies to other parts and to special places in the world, like the vortices that Imelda experiences swirling throughout herself and radiating out into other people.

These early pioneers rejected the traditional notion that the images are illusory or "subjective" in contrast to the "true" or "objective" measurements. Such images describe the bodies more fully than do the simple measurements; they tell more truth. In Sister Ruth's case, the distortion of her hips and shoulders in her drawing tells of how she has chronically contracted the muscles surrounding her hips to suppress her sexual desires. On the other hand, she always felt comfortable using her shoulders to lift patients and do the work needed in her convent. After she took up an exercise program and began to lose weight, her measurable body moved in the direction of her drawing, her hips becoming slimmer and her shoulders broader. Imelda's drawing describes her unusual perceptions of the world, making sense for us of how she moves and chooses her life goals.

A photograph or set of measurements describes the present status of Descartes's "extended thing." A neurological description gives an account of predictable behaviors and ranges of possible changes. But the image tells us about the *bottich*, a life process. Each of us is both (1) a publicly measurable object which weighs so much, whose dimensions are such and such, with a specific genetic constitution; and (2) a self-perceived body which is in the process of becoming heavier or thinner, shorter or taller, in relation to a uniquely experienced life in the world.

The gaps between (1) and (2) yield fruitful secrets. An anorexic teenager typically experiences herself as too fat even when she weighs significantly less than normal. As she explores that disparity she may discover troubled feelings about her parents' divorce, their fights around the dinner table, arguments about the proper kinds of food to eat, and the dangers of having a sexually attractive body. A well-muscled, heavy-boned man may experience himself as fragile. With the help of a therapist in exploring the gap, he may find that his mother encouraged him to exaggerate his illnesses as a child so that she could keep him from his father's world.

Pause

Do you think of yourself as heavy or light? Strong or weak? Tall or short? Beautiful, handsome, unattractive, or downright repulsive? Are there discrepancies between these feelings and what you know from scales, measurements, and people's feedback?

Would you answer these questions differently from day to day, even though you know you are not changing very much? What provokes such changes in your self-esteem?

Freud further eroded the identification of "body" with *corpus*. He rudely showed that Descartes's *res extensa* is simply the way one organizes one's body to defend against a world perceived as physically threatening. The helpless infant fears being devoured by its mother or destroyed by its angry father. Adults are threatened by the hardships of physical existence and the horror of impending death. To cope with these fears, we create our ideologies, wanting to believe that they are based on divine inspiration or scientific induction. Under the sway of these delusions, we retreat into the inner world of the soul, pretending that the flesh is not real or has no ultimate significance.[9] The split between a supposedly intellectual, spiritual self and a material *corpus* is a basic instance of repression trying to make itself look respectable.

For Freud, "my body" is like a dream in that its parts and their organization are the actors in a drama. The hand is not just the reality outlined in an atlas of anatomy; it is also that which I must keep away from my genitals if I am to avoid hell, that with which I seize the meaning of my life, the source of healing and blessing. The heart is more than a pump, the spleen more than a lymph factory. Tics, blushing, limps, illnesses, and stuttering are not failures in a machine but expressions of the deepest secrets of the human soul.[10]

We now know, thanks to Freud, that "my body" is not simply that publicly observable thing which weighs 140 pounds on an accurate scale, is five feet nine inches tall, fits into a size 39 jacket, and has guts and lungs where anatomy texts indicate. My father may be crouching in my shoulders, my mother lurking in my belly, Satan in my anus. Trees can grow in my spine, dense weeds in my chest.

Freud's pupil Wilhelm Reich elaborated on Freud's early interests in the neuromuscular basis of psychology. Reich originally came to his insights in the course of a seminar on sexuality in which he and other psychoanalysts tried to understand sexual dysfunctions in their patients. He found that so-called "mental illnesses" were always accompanied by disturbances in the ability to experience orgasm. In patients who began to experience more orgastic satisfaction, neurotic symptoms began to clear up. But the difficulties with orgasm were not simply "mental"; they involved constrictions in breathing, tightening of the abdominal muscles, and even squinting of the eyes. Reich also noticed the similarities between characteristic rigidities in a person's ordinary body movement and that same person's resistance to the pleasure of orgasm. A characteristic posture, such as lifting up the chest and tightening the back, would become exaggerated during sex to the point where it prevented yielding to orgastic impulses.

In case after case, Reich was struck by the correspondence between muscular rigidity and intrapsychic conflicts. When he was an intern at the Steinhof asylum in Vienna, for example, he encountered a young woman with totally paralyzed arms and muscular atrophy. Neurological examination could reveal no organic disorders. She told him that a shock had precipitated her paralysis: her fiancé had wanted to embrace her. Half terrified, she stretched her arms forward "as if paralyzed." Afterward she could no longer move them. Atrophy gradually set in.[11]

During this same period, Reich was analyzing a waiter who was incapable of having an erection. The man was very quiet, well-mannered and did everything that was asked of him. He never got excited. During the course of three years of analysis, he never once became angry or critical. According to the prevailing concepts, he was a fully integrated, adjusted character with only one acute symptom. The analysis went smoothly. In the third year, they arrived at a reconstruction of the "primal scene," in which the man, when two years old, had witnessed his mother giving birth. The impression of a large bloody hole between her legs had been ingrained in his mind. On a conscious level, there remained only a sensation of an "emptiness" in his genitals.

According to the standards of the Freudian circle, Reich was judged to have completed a successful analysis; the man was thought to be well adjusted though impotent.[12] But in later years Reich discovered that such typical "well-adjusted behavior" was the result of a carefully orchestrated set of muscular rigidities designed by the person to inhibit fearful impulses or rage. As he observed people expressing that rage, he found that anger itself was made up of an even deeper set of muscular rigidities inhibiting still more dangerous feeling of love and pleasure forbidden by the culture. He called these complex networks of muscular rigidities, which permeate one's whole body, "character armor."

> I compared the stratification of the character with the stratification of geological deposits, which are also rigidified history. A conflict which is fought out at a certain age always leaves behind a trace in the person's character. This trace is revealed as a hardening of the character. It functions automatically and is difficult to eliminate. The patient does not experience it as something alien; more often than not, he is aware of it as a rigidification or as a loss of spontaneity. Every such layer of the character structure is a piece of the person's life history, preserved and active in the present in a different form. Experience showed that the old conflicts can be fairly easily reactivated through the loosening of these layers. If the layers of rigidified conflicts were especially numerous and functioned automatically, if they formed a compact, not easily penetrable unity, the patient felt them as an "armor" surrounding the living organism. This armor could lie on the "surface" or in the "depth," could be "as soft as a sponge" or "as hard as a rock." Its function in every case was to protect the person against unpleasurable experiences. However, it also entailed a reduction in the organism's capacity for pleasure.[13]

For example, Sister Ruth explained her life experience in Reichian categories. She experienced a hardening of connective tissues around her genitals and around the areas of her surgery. That hardening kept her from feeling the desires forbidden by her religious vows and protected her from experiencing her intense back pain. She made her rib cage rigid to restrict breathing.

Reich made a significant move beyond Freud and other psychiatrists into the social context of bodily repression. He saw that crowded homes, poor working conditions, the oppressed role of women, and the lack of birth-control information made it impossible to have a healthy sexual life. Furthermore, the rise of fascism in the late twenties, supported by German Christians, made it clear to Reich that authoritarian ideologies capitalize on the control of bodily impulses. Fascism, Christianity, communism, and capitalism require citizens whose flesh has been rendered passive, armored bodies resisting pleasure and hungry for charismatic authorities who might fill feelings of emptiness in their genitals.

FLESH AND BONE: THE SOIL OF BELIEFS

Those early discoveries of scientists exploring the territory between neurology and psychology eventually led to a radical revision of the old photographic notion of perception. At this stage of history, we are in a better position to understand how our bodies, with their unique perceptions, figure in the construction of our more abstract belief systems. Until recent years, belief systems seemed to have a life of their own, revealed by divinities or commanded by logic and independent of the geographies and physiognomies in which they were born. But we are coming to have an increasing sense of how those ideologies are rooted in neuromuscular structures formed in tangible milieus.

Even though you and I have skin, eyes, tongues, ears, and noses which behave according to similar neurophysiological patterns, we perceive the world differently. What we see when we look at a forest is not exactly the same. We organize space and color in different ways. You may focus more on the smells of the world, I on its textures. You may feel more comfortable within the small spaces of enclosed gardens, if you are more sensitive to the intricacies of the tiny landscapes under branches. I am more at home in open vistas, having a sense of the pattern of mountain ranges but not much feeling for detail. The people whose presence often excites you may leave me limp.

Because both of us cannot occupy precisely the same

place at the same time, and because our sensing bodies are shaped so differently by our unique histories, the ways in which you and I organize our worlds will be at least slightly different. My values and philosophy of life have been distilled from half a century's experiences confined mostly to the Sacramento Valley, the Pacific Coast, and the Southwest, seasoned with periodic immersions in New England and Western Europe. I've avoided feeling what it's like to clash in contact sports like football and hockey, or to push the limits of physical danger in hang-gliding or river-rafting. I've never seen an arctic tundra or a tropical jungle. I've never smelled a Middle Eastern bazaar. The closest I've come to sensually touching the skin of a black person or an Asian are formal handshakes and bumps on public transportation. I've organized my treasury of perceptions, with their gut reactions and muscular tensions, into a world view, in response to my parents, teachers, and friends. You have done it differently.

In the realm of values, our unique bodily inclinations move us one way rather than another. The tasks we pursue, what we choose to study, and the skills we develop are functions of our personal bodily tendencies. Sometimes our choices are motivated by relief from pain. At other times we are pursuing sexual attraction or plotting ways of protecting ourselves from having to deal with it. Sometimes it is bodily fear that motivates us; often, the reality of death.

Sexual characteristics sometimes become a major determinant of one's more abstract viewpoint. We men, with exterior genitals subject to obvious and dramatic change in a matter of seconds, often plunge through life crudely inserting ourselves into whatever spaces satisfy our urges. Only minimally aware of slower and more subtle body rhythms, we have readily taken to lives of abstract withdrawal from the organic world. A woman's body does not manifest such rapid and extreme changes in sexual organs; the buildup of sexual excitement and orgasmic release are generally slower and longer-lasting. A woman's body is in a state of more constant change, including menstruation and pregnancy. Sexual differences have of course been exaggerated into the stereotypes and oppressive social policies associated with biological determinism, but the truth in the stereotypes is that different kinds of bodies give us different raw materials for artfully constructing our viewpoints on life.

Physical malfunctioning can often provide a basis for the development of uniquely human achievement. Some of the leading creators of the new body therapies, whose lives I discuss in subsequent chapters, were people such as F. Matthias Alexander and Elsa Gindler, who had maladies for which contemporary medicine had no cures. Not accepting traditional diagnoses, they experimented until they discovered how to heal themselves. Along the way they discovered new methods for working with the body which they eventually taught to scores of others.

People born with structural anomalies or serious illnesses, such as cerebral palsy or polio, often develop a perspective on life that leads them to notice aspects of reality that pass by the rest of us. The Indians of the Southwest give a privileged place in their rituals to such people, considering them messengers from the gods.

The most abstract products of human history — philosophy, theology, mathematics, and science — are derived from bodily perception. In many cases, they represent people's conscious attempts to break out of their privacy.

Platonic philosophy, for example, was born of Socrates' attempts to refute the Athenian Sophists' arguments that our perceptions are so private that there can never be any generalized standards for truth and value. Justice, they argued, is what the powerful dictate. Plato attempted to show that we do indeed sense that there are ways to communicate that transcend our particular tastes and feelings. Even though he fathered European dualism, he gave compelling analyses of the ways in which truth and value are grounded in perception. In his philosophy of education, he argued that a man's sensibilities needed to be refined by gymnastics and music before he could learn the more abstract mathematical and philosophical realities. According to Plato, the soul's highest attainment, the unitive vision of the Good, comes about as a result of an impulse set in motion by the very sensual attraction of one body for another — like aging Socrates, transported into idealistic bliss by his passion for beautiful boys. Plato's ideal Republic was constructed on the paradigm of the perfectly functioning human organism.

Saint Paul, hardly a champion of the body, nevertheless used it as a metaphor for the true church when trying to teach early Christian communities how to heal their divisions. "The Mystical Body," based on a well-functioning human body, with many parts working towards a single goal, was his central image for teaching contentious believers how to transcend their individual proclivities. Saint Paul's notion functioned like Euclid's theorems to give people a sense of how to contribute their private vantage point to a unified arena of action.

Even the seemingly impersonal fields of geometry and physics were born of attempts to transcend idiosyncratic perspectives. Standing in Mexico City on June 21, one sees the sun describe a certain arc in the sky. Someone else at the North Cape of Norway on that same day sees the sun describe a different path. From at least the time of Stonehenge and Chaco Canyon, astronomers have tried to devise ways of determining the relative movements of sun and earth independently of where and when one stands to observe.

Galileo and Descartes rightly noted that instruments and mathematical language can resolve people's disagreement about the qualities of what they are perceiving. Special relativity and quantum theory are ways of describing the largest and smallest aspects of the cosmos in a mathematical language developed expressly to transcend the peculiar coordinates of the observer. Their unique success in comparison to the naive objectivism that preceded them has been due partly to the fact that they take systematic account of the relation between the position of the bodily observer and the observed. We can compare the discoveries of people such as Einstein and Heisenberg to the way in which we only make sense of a photograph by knowing that it was taken by a person using a certain kind of camera in a specific place. Without that knowledge, we see the flat image in a different way.

Mathematics, the most abstract of human products, is grounded in human flesh. Theorists inquiring into the nature of number are driven back to human perception, investigating what it means to count, to grasp the difference between "one" and "two" or the meaning of a series.

Even religious ideologies that one would expect to be rigidly dualistic regularly present their notions of the divine in the images and language connected with a particular culture's sensual experience, despite theologians' constant disclaimers that these concepts truly represent the divine. The Homeric gods and goddesses, Hindu avatars, Hopi kachinas, and the Christian trinity are clothed in human forms peculiar to certain historical times and places. Despite Judaism's and Islam's interdiction of divine images, their language reveals notions of *him* as a male body. More abstract expressions of the divine, such as one finds on Tibetan mandalas or Islamic mosques, reflect patterns of physiological energy flowing through the body.

I concluded the last chapter with an analysis of the perversion of values that moves our social systems. We are now in a better place to understand the way in which that perversion operates. Individuals within a particular belief system forget that their particular ideals and values grew out of a specific plot of soil. They cling to their beliefs as objective statements of the way things "really are." In that forgetfulness, one group feels compelled to convince others of its truth, even if force, or, in the ultimate instance, destruction is needed.

We have organized ourselves socially so that we think there is something amiss when, for example, Sister Ruth and Imelda draw such radically different pictures of their bodies. One must be more right than another, by comparison with some ideal "right." We have lost a sense that the ideologies within which we live evolved from the bodies in a particular community adapting to cope with a unique climate, feeding on certain types of food, and being shaped according to particular languages and music. Instead of relating to our ideology as derived from standing with a particular style in certain geographical locations at a given historical moment, we act as if it fell from heaven as Truth, to be defended to the death. Too busy with asserting that Truth, we have no time for more basic issues, like eliminating world famine.

To think of the world as a community of bodies, *bottichs*, rather than as a collection of *corpora* is to think of various ideologies like the spirits one finds in different regions of the world: grappa in Italy, armagnac in Languedoc, sourmash in Tennessee, sake in Japan, slivovica in Yugoslavia, or samsoo in China. The different tastes and aromas of those liqueurs derive from the grains, fruits, waters, and brewing traditions of different countries. Instead of getting drunk together with tasting each other's special spirits, we are trained to become corporeal machines, painfully bouncing off each other like little bump-cars in amusement parks, powered by someone turning dials in a remote booth.

CHAPTER FOUR

THE SOCIAL BODY

Because our belief systems are distilled from our flesh, they can be watched and shaped. My parents watch me to see that I behave and punish me when I don't. Teachers, sharing the parental burden, refine the watching, making sure that I learn to be quiet and sit straight. Music teachers, coaches, and job supervisors watch to see that I perform the way they expect me to. With the help of its many agencies the State watches me. Through the eyes of my well-schooled conscience God watches me, even in the darkness of my bed, seeing if my hands tough my genitals under the sheets or if I conjure up erotic images in my brain.

Because my viewpoint can be watched and shaped, it can be coaxed, even forced, into alignment with the viewpoint of my community. My body — its sensibilities, movement styles, reaction patterns, and health — is not simply an individual reality governed by its own biophysical laws and the idiosyncratic effects of my personal history. I am also a result of the ideologies within which I move.

It may sound odd to you when I say that *belief systems* can be watched and shaped. Our immersion in dualism makes it difficult to grasp the connections between belief systems and our patterns of body movement. But we can see and hear the belief systems of a society in their native dances and music. The difference between the designs of a gothic cathedral and a Quaker meeting house is a visible difference of belief. The way in which a group of children sit in their school desks shows us a philosophy of education. When we watch Leni Riefenstahl's films of Hitler's rallies at Nuremberg, we are seeing Nazism.

Each of our bodies is an artifice, a community project visibly manifesting the values of those implicated in the task. Like a mosque, tract home, or courthouse, the adult body is the result of the community's imposition on the earth's raw materials of those forms that serve the current needs of that community. Charles's carriage reflects upper-class New England ideals of uprightness and self-control. Sister Ruth's body tells stories of Germany, American medicine, and Roman Catholicism. Imelda says she carries her shoulders like a Cherokee.

Designs of the body vary with time, the prevailing architecture of one period giving way to another. They also vary from culture to culture, as American Indian pottery differs from Japanese. Within a given community, the designs vary according to a particular person's function; a worker, for example, is shaped differently from an intellectual.

It was no surprise for me to learn that all of Sister Ruth's ancestors were German. Her well-defined and somewhat heavy gait is not unlike the patterns one sees in people walking along the canals of Hamburg. But Imelda is different. Her light footfall, her way of moving so that her passage is barely noticed in a crowd, made it hard for me to guess her multicultural roots. Since both women are from middle-class working families they do not exhibit the sophisticated posture training that Charles got at Exeter and Harvard.

Those who have best understood how to control human communities — charismatic religious and political leaders, advertising consultants, thought-control scientists — know that the body is not inevitable, a given. Its tastes can be dulled, its ocular range narrowed, its emotional reactions channeled in one direction rather than another. The supple newborn can be trained in mechanical behavior so that once grown up it will react to its world predictably and passively, buying the right products and voting for the right candidates. Shaping the flesh becomes crucially important in the organization and maintenance of power.

Pause

Consider how much you conform to objects in your environment.

When you're buying shoes, how careful are you to test them out to see how they will affect your knees, pelvis, and neck? Do you notice whether the shoe's heel falls directly under your heel? Do you test many lengths, widths, and styles? How assertive are you in the face of your salesperson's judgments and the demands of fashion?

How active is your relationship to the chair you're sitting in? Do you use sensations of discomfort as signs for adjusting your position, or do you just let yourself collapse into the countours?

Are you careful to adjust your automobile seat so the distance between you and the gas pedal is just right? So your back and head are comfortably supported? Do you make refined adjustments in your desk chair? In your airline seat?

If you are a teacher or a student, do you simply accept the way desks are arranged, or do you give some attention to how seating patterns can affect learning?

If you attend meetings, you might observe how active people are in relation to the arrangement of chairs and tables. Do they exert ingenuity in making the furniture conducive to sensitive dialogue?

Developing a clear awareness of how we relate to various molds is the key to recovering our authority. For example, as you become aware of such a simple pattern as being trained to sit still in your chair or at your desk, you may find that your body is in fact dictating subtle movements, shifts of weight. You may feel like getting up and walking periodically, or stretching for a couple of moments. As you give more attention to these sensual impulses, you may find that you feel less sluggish, or that your back is less sore at the end of a working day. You may even become dangerously critical of the way in which your work environment is engineered. You may be chided for being too restless.

"TECHNIQUES OF THE BODY"

It's not easy to grasp how radically plastic the body is, precisely because one of the popular convictions is that the body is simply a corporeal thing that is structurally altered only by deterioration through age, accident, or illness. There is, of course, a substratum upon which culture builds. Genetic patterns determine the serial relations among body parts so that fingers do not generally appear on the foot, the heart is located somewhere in the upper part of the chest, and muscular tissues appropriate for the scrotum do not appear instead in the biceps. But that substratum, governed by biophysical laws, provides an unimaginably large range of possibilities. Even the embryo is shaped by the patterns of its mother's life; her movement induces certain postures and stresses which affect, for example, the shapes and precise locations of the child's developing tendons. Obstetricians such as Lamaze and Leboyer have shown how the particular methods whereby a given community deals with the birth process will have profound effects on a baby's neuromuscular responses.

The variability of body structure throughout history and from one culture to another is made possible by the plastic character of the biological data. In contrast to the other higher mammals, the human infant's nature is more or less unspecified. At birth, for example, the human brain weighs approximately one fifth of its ultimate weight, whereas the anthropoid brain weighs two thirds. The embryological development of humans can be said to extend long after we have emerged from the womb into the social world.

Consider that the average newborn is only beginning to form the shapes of some 190 bones within a largely undifferentiated mass of connective tissue. The junctures between each of these bones offer virtually unlimited possibilities for different arrangements among the components. The manner in which the infant moves and is moved, particularly during its early years, will determine the length and shape of bones, the shapes and positions of muscles, and the relations among the various segments of the body. For example, the foot's twenty-eight bones and thirty-two joints provide only a very general outline for its eventual shape. The curve of its potential arches, the organization of the toes, and the position of the heel in relation to the lower leg will be determined by such things as how soon the child starts to walk, the kind of shoes it wears, and the different weight it places on each foot.

Even adults have numerous possibilities for change. For instance, we might feed into a computer the structural variations available to a particular forty-five-year-old woman, along with all the conceivable methods she might use during the next ten years to realize one possibility or another. One track might include daily yoga, therapeutic massage, and golf. Another might involve months in bed with a serious illness. Another might involve encountering a new and exciting lover. Examining the results of all conceivable programs, we would find a wide range of possible relationships among her head, chest, pelvis, and feet.[1]

Superimposed upon the protean infant to create the bodies we know are the culturally variable ways in which we learn to use our bodies. The French sociologist Marcel Mauss called these various ways "the techniques of the body."* He argued that even very simple activities which people

---

*Les techniques du corps would be more properly and cumbersomely translated as "the ways in which we learn to use our bodies." I will continue to use techniques as a convenient signal that ordinary activities which we often take for granted are actually methods for giving our bodies characteristic shapes. In later chapters, for example, I will contrast two "technologies" or approaches to educating our bodies, one evoking alienation and the other authenticity. I use technology in its "pre-technological" sense, from the Greek word for making things, for art — in the sense that weaving, cooking, and pottery are technologies. In this case, I use it to refer to a system for forming our bodies in specific ways.

tend to think are natural are learned from society. After World War I Mauss noticed that British soldiers had difficulty digging with French spades and marching according to the rhythms of French military bands. He noted that people from some countries learn to swim and dive with their eyes open, others with their eyes shut. He was struck by the differences in walking styles between nurses in the hospital and women attending the cinema. Even such seemingly natural activities as nursing infants and sleeping are done in different ways in different cultures. "And I concluded," Mauss wrote, "that it was not possible to have a clear idea of all these facts about running, swimming, etc., unless one introduced a triple consideration instead of a single consideration: physical, psychological, and sociological. It is the triple viewpoint, that of the 'total man,' that is needed."[2]

Perhaps without being able to articulate exactly what you've seen, you, like Mauss, may have noticed and wondered about regional variations among people with the same racial and cultural roots. We all have vague perceptions of such patterns as the swagger of Texas cowpunchers wearing Tony Lama boots, or the plodding of Vermont farmers in their Redwing clodhoppers, or the well-aligned stride of Wall Street bankers in their Johnston and Murphy brogues, or the jaunty gait of California psychologists in their Nikes. When you've been in a large airport or bus terminal, you've seen an enormous amount of data about the cultural shaping of bodies, even though you may have been educated not to notice it. In such places you often see not only regional variations in travelers from different parts of the country, but also degrees of cultural assimilation, as when you see an American welcoming relatives from another part of the world. As they walk through the terminal, you may notice not only differences of clothing and accent, but also differences in the way they take strides, move hips and shoulders, and carry their heads.

Evidence for the difference between genetic structure and the results of a society's techniques imposed on that structure is found in large cities of the United States, where peoples from traditional cultures are in various stages of being blended into the American stew. In San Francisco and Los Angeles, for example, there are many fourth-generation Americans of Japanese descent. If you were to mask them to hide such racial features as the color of hair and eyes, skin texture, and the shape of the skull, you could probably not distinguish them from fourth-generation Americans of Irish, Polish, or African descent. These people all share the same kinds of muscular development, speech inflections, walking styles, and ways of gesturing. By contrast, tourists visiting from Japan have unique patterns of body organization; they do not move like tourists from Senegal or Russia.

The minimal truth underlying our racial stereotypes is that different cultures do have different styles of body use and development. Traces of traditional styles are often preserved within ethnic communities. In Americans of Latin origin, one often sees the patterns of movement associated with dances popular in the Caribbean, which have a great deal of movement in the hips in relation to the trunk, which is held erect. Some black Americans still move with the clear differentiation of body parts that characterizes African dance; they still have an ability to move the trunk with a high degree of independence from movements of the head and pelvis.

The vast number of techniques found in any society are taught in different ways. Some are transmitted through suggestion, by the ways in which adults hold and touch their infants. Others are taught by example; children learn to do basic activities simply by mimicking adults. Still others are communicated through the design of the environment: styles of clothes, kinds of furniture, and designs of dwelling that evoke certain patterns of bodily development. Art and imagery are powerful vehicles for the techniques. Explicit instruction is possibly the least significant mode of teaching.

The point of the following survey of techniques is to give you an idea of where to look in your own history for the molds that have produced your present behavior. As you develop a more accurate feeling for the contours of these molds, you may find their hold on your movements less tenacious.

TECHNIQUES OF INFANCY AND CHILDHOOD

Infant wrappings, diapers, cradles, slings, walkers, and other paraphernalia determine patterns of muscular development. We can understand the culturally unique aspects of customs that may seem natural to us more clearly when we contrast them with customs in other societies.

The Navajo, for example, strap their infants to cradleboards, while the Balinese carry their babies loosely on their hips or in slings. The straight lines of the boards are similar in pattern to the art forms and body structures of Navajo adults.

By contrast, the looseness of Balinese infant-carrying practices is reflected in the flexible extended bodies of Balinese dancers. Margaret Mead wrote of the Balinese sling that it

> permits the child to be attached to the mother or the nurse without either person's making any active effort whatsoever once the sling is fastened, and when the sling is absent, the Balinese arm imitates it in relaxed inattention. This very light tie between child and carrier, tactually close, but without grasping by either one, seems to establish a kind of communication in which peripheral responsiveness predominates over grasping behavior of purposeful holding on.[3]

That "peripheral responsiveness" is used to advantage by masters of music and dance, who evoke the appropriate postures of their students by light touch.

Many societies shape their infants at an age when their bones are extremely malleable. Among Indian tribes such as the Maya it was common to bind the skulls of selected infants between two pieces of wood so that their heads would acquire the elongated shape apparently associated with high status. The Kazaks in the Tashkent area of the USSR customarily elongate the heads of newborns by using sandbags. When the practice fell into disuse, the local government ordered it reinstated to distinguish their citizens from nomadic Turks, who were becoming more numerous in the area.[4] West Indian mothers practice a sophisticated form of massage with their infants in which they mold the baby's head, nose, small of the back, and buttocks.[5] Europeans have used head bindings to create elongated skulls.[6]

Figure 3

55. French child wearing a 'bandeau' (after Foville).

56. 'Deformed' head due to bandaging. (Dingwall)

57. 'Deformed' head due to use of tight bands (after Foville).

58. Diagram showing a 'deformed' skull. (Imbelloni)

An infant is taught to crawl by the way in which adults allow it to move. Even such a seemingly "natural" activity as sleeping affects body shapes. The design of the bed, the thickness of the mattress, the kind of wrapping used, and the ways in which the parents place the infant will evoke different kinds of shapes and muscular tensions in the body.

Something as banal as diapers can have significant effects on structure. A large plastic diaper used regularly for several months will coax the child's legs to be slightly more everted than if it wore less bulky, softer diapers. As the child grows up, he or she might have short gluteal and rotator muscles of the hips and overextended adductor muscles.

Walking is also a major determinant of the shapes and

interrelationships of a child's bones. Pressures to walk at a certain age, ways in which the child is supported, and the ordinary movement patterns of its parents give necessary lessons for this skill. The child will have to adjust head, torso, and hips to maintain balance on feet and legs with slightly different shapes and sizes. Those compensations will give rise to relatively permanent torsions throughout the musculo-skeletal system.

Learning one's native language is a primary factor in giving the body its unique cultural shape. The proper pronunciation of the mother tongue determines the relative strength of the muscles of one's mouth and the structure of the oral cavity to such an extent that it is often difficult to adjust to the inflections of another language. Besides shaping the mouth, face, and throat, learning one's language also shapes the body by requiring a distinct gestural vocabulary with characteristic shrugs and postures.

Like language, one's home constantly evokes specific patterns of bodily development. The heights of its ceilings and the size and shape of rooms will suggest certain styles of body movement. The relative location and design of the bathroom, bedroom, and dining room suggest lessons about defecation, sex and eating. Entries, yards, window coverings, and hallways will teach about bodily orifices and privacy.

The architecture of one's village or city will expand the environmental lessons of the home. The geometrical, traffic-filled streets of Manhattan, lined with skyscrapers, encourage a child to develop certain perceptual and motor skills adapted for survival in that distinctive environment. The open farm-lands of Kansas will evoke other perceptions and controls; the narrow, twisting streets of the Zuñi pueblo still others.

The techniques of eating are a major expression of how a particular society structures its control of the body. European and North American children are taught to eat their food while sitting in chairs in front of a table. This already requires a unique form of skeletal arrangement. Add to that the neuromuscular skills necessary to manipulate spoon, fork, and knife. The child must also learn the refinements of social behavior associated with eating: the proper speed, acceptable ways of opening and closing the mouth, the kinds of controls to be exercised over chewing, belching, and farting, the restraints on interacting with others at the table. The complex of these rituals requires a highly developed and unique muscular usage that produces a unique body structure.

In India, by contrast, children are taught to eat with their right hand while sitting on the floor. That requires, among other things, a hip structure in which the thighbone is capable of more eversion than is required in the Western style. It also teaches a different sort of manual dexterity from that required for European utensils. The Japanese child also sits on the floor but at a low table; he or she must learn to use chopsticks. When adult Westerners try to adopt these techniques in Oriental restaurants, zendos, or ashrams, they often experience discomfort and awkwardness, since they are unprepared by the early shaping of bones that occurs in children.

The rituals surrounding urination and defecation constitute another important group of muscular lessons. Sitting on toilets in private rooms is an integral part of life in modern culture. This requires sophisticated control of the sphincter muscles to restrain the flow of urine and excrement during the sometimes long periods one has to wait before reaching the private shrine.

Even such seemingly spontaneous activities as sneezing, coughing, and vomiting are shaped by society. Vomiting, for example, is commonly thought to be disgusting. Children are subtly taught to keep such activities out of the public sphere as much as possible. But within the peyote ritual of the Native American Church, vomiting is honored as a divine blessing. Among Reichians and bioenergetic therapists, vomiting is regarded as an effective way of dissolving body armor.

Already, in these fundamental activities of human life, one can discern basic patterns of social organization: attitudes toward emotional expression, modes of social intercourse, the manner of relating to authority figures, and the significance given to rudimentary bodily processes. These patterns are gradually etched into infants' bones, muscles, and neural pathways.

SEX

One of the most important set of techniques for anchoring authority is that which differentiates the bodies of men and women. Boys learn the swaggers and shrugs appropriate for men, including the nonverbal vocabulary that signals their socially assigned superiority to women. Girls learn the facial movements, ways of holding their hands, and postures that manifest their roles as adjuncts to men. They also have to learn their culture's particular ways of dealing with menstruation, pregnancy, and childbirth.

In addition to the almost imperceptible milieu of sex-dividing techniques are specialized and different ways of shaping boys' and girls' bodies. Pressures will be put on boys, for example, to develop their large shoulder and thigh muscles in activities like playing football and climbing trees. Often boys' bodies will be even more muscularly shaped by lifting weights or by training in a martial art such as boxing or judo.

History has produced an overwhelming number of female-shaping devices, which should give important clues about humanity's fears about women's bodies, the subject of Chapter Seven. Some of these techniques are extreme and obvious. Chinese foot-binding had the aim of both controlling women's movements and piquing men's sexual appetites. In Africa and the Middle East, women's labia are sewn together and their clitorises excised to keep them from yielding to sexual passion. A recent, conservative estimate by the World Health Organization puts the number of women who are so mutilated today at some 30 million. A representative of that organization says that "the purpose is to reduce or extinguish women's sexual pleasure and to keep women under male sexual control. African and Moslem men often refuse to marry girls who have not been operated on."[7]

It's easy for Westerners to think of these techniques as primitive and barbaric without noticing our own practices of "aesthetic" surgery: breast and buttock lifts, surgical implants, nose jobs, and facial sculpturing. While these techniques seem more dignified because they are performed in physicians' offices or modern hospitals by highly paid surgeons, they often leave women scarred and in pain. Some such surgeries result in death. Even in cases where the surgeries are successful, the message is the same as the one in African villages: women are shaped to please men.

Besides these extreme techniques are a wide range of body-shaping devices for women. Western Europe and the United States have developed bodices, corsets, and breast-shaping bras. Women in African and Asian cultures often use necklaces, earrings, and lipdisks to alter their appearance.

Even sexuality, which is popularly thought to be the most spontaneous of activities, involves a highly elaborated set of techniques which vary from culture to culture. Courtship rituals include how to signal attraction, the proper way to make the approach, and how to receive or spurn it in ways that keep the excitement building. Once the woman and man make contact, they demonstrate affection by socially accepted ways of caressing and kissing.

Sexual intercourse involves a wide range of what are considered appropriate techniques involving postures, intervals, and rhythms for lovemaking. Such techniques range from those sophisticated, complex postures that are well known from Oriental art to the simpler postures illustrated in Western pornography. In the Orient, the techniques are theoretically elaborated as a part of the methods for developing a spiritual practice. In the West, the theoretical study of sexuality is primarily confined to medicine and psychology.

Pause

What kinds of gestures and postures do you consider effeminate? macho? Can you recall incidents in your childhood when your mother or father told you that certain ways of using your body were (if you are a man) not manly, or (if a woman) not feminine?

What has it been like for you to experience sexual stimulation, particularly when you are in public places or with strangers?

Recall attitudes you picked up within your childhood home towards sexual fluids: semen on the sheets, menstrual flow, vaginal secretions.

Imagine the range of positions you use in making love. Think of the ones you prefer. Are there some you hesitate to try? Do you allow yourself to explore wide ranges of activity and passivity in making love?

When you feel orgastic impulses, do you allow them to move freely throughout your body, or do you restrict them by tensing specific muscles or slowing your breath?

WORK

The techniques demanded by work will have a significant impact on shaping male and female adults. The body structure required to manipulate a Chinese handplow differs from that required to run a John Deere caterpillar. A life of washing clothes on stones by a river and hauling wood for cooking requires heavy musculature that is not needed to use laundromats and modern kitchens. Many of us have to develop highly specialized motor skills to manipulate typewriters and computers, and to learn how to sit relatively still for long periods of time. Surgeons develop the skill of manipulating the scalpel, dentists the drill. Some people, such as homemakers and farmers, need a wide range of muscular abilities. They have to have strength in their large muscles to do heavy lifting as well as more refined skills for such things as sewing and repairing equipment. Most of us have to learn the sophisticated neuromuscular patterns required to manipulate a pencil and drive an automobile.

Karl Marx was one of the earliest thinkers to notice how one's life work shapes one's body and perceptions. A person's body often begins to take on the nature of his or her tools.

"The sense caught up in crude practical need," Marx wrote,

> has only a restricted sense. For the starving man, it is not the human form of food that exists, but only its abstract being as food. It could just as well be there in its crudest form, and it would be impossible to say wherein this feeding activity differs from that of animals.

> The care-burdened man in need has no sense for the finest play; the dealer in minerals sees only the commercial value but not the beauty and the unique nature of the mineral . . .[8]

A worker on the assembly line or a secretary in an office, performing only one kind of action throughout the day, begin to get a sense of their bodies as machines with a narrow range of movement and little feeling. The reduction of the body's capacities to the specific range required by habitual work correspondingly diminishes the scope of one's perceptions.

A TECHNOLOGY OF ALIENATION

It would be misleading to consider the various techniques of the body as discrete things. Within any human life they form an organic pattern. Learning to play golf and to dance are elaborations of the way in which one has learned more basic activities such as walking and grasping. The different ways in which mothers and fathers touch (or refrain from touching) their infants will influence the children's patterns of moving and responding as men and women. All these specific activities are related to how the society thinks its members should control their emotions and sexual feelings.

Wilhelm Reich's notion of character armor represents his understanding of the organic unity of the techniques. He first began to understand the notion when he observed victims of an epidemic doing calisthenics. People did the same exercise in different ways, and the differences were stereotypical: a professor did them in a "professorial" way, a convicted criminal in a "criminal" way, a hyperactive adolescent in a "hyper" way.[9]

Failure to grasp the relationships among the techniques is related to our inability to design effective social reform. Obstetrical procedures, teaching sports, and stress-reduction techniques are commonly evaluated as if they were separate entities rather than organic parts of a social system. New techniques are often developed to replace those that seem ineffective, but the social goals remain unexamined.

Throughout the remainder of this book I will be contrasting two ways of integrating the techniques. I call one the "technology of alienation," a way of applying the techniques that trains people to be disconnected from their sensual authority; I call the other "the technology of authenticity," a way that encourages people to develop their own sense of authority, their peculiar genius.

The contrast is apparent in the case of a two-year-old infant I will call Johnny. Johnny had a characteristically infantile pattern of bowed legs and everted feet. Concerned that he would grow up with that shape, his mother took him to an orthopedist, who prescribed a brace that would force his knees and ankles into a straight line. He told her to put the braces on Johnny at night. Not surprisingly, the child screamed all night, to such an extent that his mother could no longer bring herself to put the braces on him. She came to me furtively, unwilling to tell her orthopedist that she was not obeying his advice.

I worked with Johnny for several weeks, simply massaging certain areas of muscular tissue to stimulate their incipient activity, and showing his mother how to do the same. When she took him back for a scheduled visit to the orthopedist, she was told that the braces had done their job: his legs were so straight that the braces were no longer necessary.

I did not straighten that child's legs, but simply encouraged him to feel more. My actions were designed to teach him that it was just fine for him to explore the full range of movements he needed to figure out how to get his feet and knees solidly under himself in a unique pattern that neither I nor an orthopedist could predict. Like Johnny, each of us masters any particular activity, from eating to playing the piano to performing surgery, by prolonged repetition of specific muscular patterns. By developing some muscles and leaving others partially or virtually totally unused, the repetition slowly determines the relatively stable relationships among the large segments of our bodies: a characteristic tilt of the head in relation to a slant in the shoulders, a turn in the pelvis manifested in a characteristic gait, a slight compression in the chest or a curvature in the back.

But within an authoritarian atmosphere these repetitions carry seeds of corruption. The practice necessary for skill is done in a context that rewards conformity to someone else's ideas rather than finding one's own way. The infant is permitted its brief moment of trial and error before learning that there are right and wrong ways of doing things. Doing things "right" becomes more important than learning.

The problem lies in the manner in which repetition is taught. Children learn that there are patterns "out there" to which they must conform: the shape dictated by a leg brace or orthopedic shoes, the correct arc for swinging a baseball bat, the pattern of steps in a dance. Early in life they learn to become disconnected from their own experiences and to direct their attention outside themselves to what authorities tell them about the proper way to use themselves. They are seduced into forgetting their native capacities for learning new ways of responding to the ever new stimuli they encounter in the sensual world. They are reduced to neuromuscular machines. The naturally experimental infant is slowly educated into becoming an adult who is afraid of trying anything new; its malleability is transformed into a structure that repeats, a character structure. A variety of methods useful for learning how to live and work together are transformed into a technology of alienation.

It is crucial to an understanding of this book to grasp the difference between any particular physical activity (the technique) and the way in which we learn to integrate it into our lives (the technology). Any technique — holding a baby, typing, driving, bowling — is of neutral value. The nature of the technology makes it a means of drawing us further away from ourselves or intensifying our sense of connection. "Fitness" techniques, for example, can be of obvious value for our health and vitality. They can be of special value to women in redressing some of the power imbalances with men and enabling them to protect themselves. But such useful techniques can easily be transformed into methods for increasing stress — sophisticated forms of self-punishment. If they are performed mechanistically, they will augment tension and reduce a person's range of movement. Such fruitful methods sometimes become like the penitential practices of medieval monks, self-inflicted penalties for overeating or prolonged sitting.

Pause

Consider the technique of sitting. Earlier I invited you to examine how active you are in relation to various kinds of seats. Recall now key incidents in your education for sitting. Think of scenes in first grade when you were introduced to hard desks arranged in rows. Remember how it felt to make the transition from preschool freedom to the regimentation of elementary school.

Do you recall being told to sit up straight or sit still, either at school or at home?

Imagine what it looked like when you were a child sitting at the dinner table. Recall how your parents would sit reading, conversing, or watching television.

Recall any posture training you had as you grew older.

As you now sit in this chair, have you given your body up to the shapes dictated by furniture designers, or to the postures prescribed by one or another expert?

Imagine how you might shape yourself for a particular goal in sitting: relaxation at the end of a long day's work, sensitivity to a friend in an intimate conversation, alertness with comfort at work, perceptiveness while driving in heavy traffic. You might try several kinds of seating and different ranges of postures in each. Notice the body states you evoke, ranging from extreme alertness to drowsiness, from pain to ease.

I invite you to experiment with the technology of sitting while you're reading this book. Often shift your position, rotate your pelvis, reposition your center of gravity, stand for a while, try a different seat. Observe what happens to your ability to concentrate and your comfort.

You might naturally apply similar reflections to any physical activity you do regularly: jogging, a team sport, yoga, driving, ironing, and so on. Simply vary the pace and form. Observe the different results.

I antitipate that through reading this book you may become more disengaged from some of the mechanical patterns that you notice in various of your activities. For example, you might see a certain grimness in your behavior at work, which is reflected in a tightness in the muscles of your shoulders and neck. The same thing happens when you are driving in heavy traffic. When you're jogging or dancing, you might find remnants of that doggedness. It may even appear when you're making love. As you become more specifically aware of the muscles associated with that pattern of your life, you will notice that they begin to relax; you can literally give more of yourself to what you're now doing and stop obstructing yourself with unnecessary muscular rigidities.

For twenty years Sister Ruth tried various techniques to relieve her back pain, without success. All those techniques were applied within a world that put great stock on mastery of so-called unruly impulses. Rigidity and self-control were always prime virtues in Sister Ruth's education, which trained her to feel that obedience to male religious and medical experts would free her from the illusions arising from her too-human passions.

It was similar with my history. I tried technique after technique in attempting to embody my adolescent insights into what it might be like to be free and to love. But I kept repeating myself, hypnotized by the voices of traditional and avant-garde "orthopedists," unable to discover my own voice.

No matter how different the techniques we use to improve our bodily life, we tend to integrate each of them into the authoritarian web that was wrapped around us during infancy. As long as we remain snared within that web, we continue to think of our bodies as disconnected, not really our own. We look to others to teach us what to do: priests, gurus, physicians, psychotherapists, accountants, and politicians. At a fundamental level of our being we think of ourselves as powerless, and regard the destructive dynamics of our social world as unalterable.

CHAPTER FIVE

IDEAL BODIES

Pressures to shape ourselves according to external images disconnect us from our own experience. Preoccupation with these images — the "proper" way to sit, the "correct" way to move our feet while dancing or playing a sport, the ideal alignment of the spine, the sexiest shape — distracts us from information our own bodies give us about how to do any activity most efficiently.

Children and young adolescents are trained in accordance with a very general notion of "straightness." That vague notion may be made more precise for those whose parents take them to an orthopedist. Braces or special shoes communicate to a child like Johnny that his right leg is too everted or his arch too low.

As we grow into adulthood and learn to deal with certain problematic areas in our life, most of us develop a more specific ideal for our bodies. For some a more precise ideal comes through seeking out various professionals to achieve pain relief. For instance, a person who often visits a chiropractor might be taught that his third cervical and fifth thoracic vertebrae are out of line and that his left leg is regularly a half-inch shorter than his right. Someone visiting a body-oriented psychotherapist might be educated to notice the curve in her upper back, which supposedly reflects a burden of repressed anger at her parents, or her locked knees, manifesting her insecurity in the world. As people try to deal with their sexual problems, they might be told to notice how "out of line" their pelvises are, supposedly blocking sexual feeling.

Being made aware of the possibilities offered by aesthetic surgery can also make a person's ideal more precise: the buttocks could be rounder, the breasts higher, the stomach flatter, the chin dimpled, the nose smaller. Orthopedic surgery can even change shoulders by cutting certain tendons, or straighten one's lumbar spine by fusing the "misaligned" vertebrae.

Popular somatic ideals function even in the lives of those who don't or can't conform to them. A tall, large-boned teenage girl with full round breasts who does not look like models in *Seventeen* begins to stoop, to cover her body with loose clothing, and to develop bizarre dietary habits. When she reaches middle age she begins a series of aesthetic surgeries to remove the wrinkles that she never sees on actresses or models of her age. An adolescent boy who is shorter than average may, after years of trying to make his body larger and more muscular, turn in despair to a life of scholarship and celibacy.

I was pressured by increasingly specific pictures of "straightness" for most of my life. One rainy day when I was a sophomore at Christian Brothers High School, our physical education instructor lined us up and told us to press our backs against the wall. He said that we should be able to feel no spaces when we slid our hands behind our backs. No matter how hard I pressed, I couldn't get rid of spaces in my lower back and neck. From that day on I often caught my reflection as I passed by a window and noticed with discouragement curves in my lower and upper back. At parties I would sometimes notice myself in a mirror and, feeling I was unattractive, pull myself up. When I entered the Jesuits I found that Saint Ignatius Loyola's "Rules for Modesty" enjoined straightness of bearing as a mark of the ideal spiritual person, so I put even more effort into making my spine erect as I walked about the seminary hallways in my soutane or knelt in meditation. But by the time I was thirty, my height was reduced by an inch and I had begun to experience chronic back pains which were so severe that I would have to get out of bed in the middle of the night to sit in a chair until the spasms abated. Around this period, the new "human potential movement" made me aware that the curves in my spine could indicate psychological as well as physical and spiritual aberrations. I learned of Dr. Ida Rolf, creator of a new form of body therapy. One of her students worked on me; marvelously, my chronic back pain disappeared and, more significant for me, I felt taller and more graceful. When I looked in the mirror I actually seemed straighter — but I painfully learned under Dr. Rolf's keen gaze that I was far from straight. She had more than a vague notion of "straightness," since she had developed a very precise notion of what our spines should look like, expressed in this drawing, the logo of her Institute; inspired by her successful work with a boy afflicted by cerebral palsy.

For her, each joint of the body (of *everyone's* body) has a precisely defined ideal position. Insofar as an individual deviates from these positions, he or she is just that much short of his or her potential. With far more astuteness than my high-school gym instructor, Dr. Rolf pointed out to me just how far I was from conforming to her ideal, which was so far that there were times that she complained that I could never learn her system.

I then embarked upon a ten-year program to get myself more accurately aligned.

Figure 4

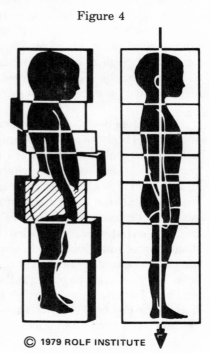

© 1979 ROLF INSTITUTE

I had my body painfully manipulated, did exquisitely balanced exercises, and attempted to build straightness into all the movements of my life, from walking to lovemaking. But the chronic pain returned with a vengeance. While I was living in Paris in 1979–80, the pain in my right shoulder was so intense that I couldn't lift my arm more than a few inches or sleep on my right side. In my various efforts to relieve that pain, I began to realize that my efforts to become straight had led my spine to become increasingly more rigid, to the point where there is virtually no movement in the vertebrae of my neck and lower back. That rigidity, reflected in my mental and psychological attitudes, was the source of my pain.

I always had an accurate intuition that it was useful for me to use my full height, not always giving in to tendencies to slump when I was depressed or tired. But that intuition became distorted into a concern about how I appeared to others, rather than how I felt. These experiences helped me understand the kinds of physical pressures that are exerted on us to capitalize on our feelings of inadequacy. Our sense that we can be more than we are is transformed into a dependence on external shapes designed by someone else.

An "ideal body" is a specifically visual design of how *the* (not "my") body *should* (not "might" or "could") look. There are many ideals, ranging from the Barbie doll to the iron-pumper. We see them everywhere. They appear in the media, in the form of both advertising models and superstars. They are in sculptures and paintings. Some of them hang on the walls of classrooms and doctors' offices. Their presence is so pervasive that we rarely notice their effects on our behavior. They tell us how much we should weigh, where our heads and shoulders belong, what the contours of our bodies should look like. They train us to be sexually attracted by certain forms and repulsed by others. Under their power, we feel good or bad about our own shapes. In some cases, a particular ideal is associated with "being good" or "being spiritual"; failure to conform has moral or religious significance.

The ideals exist outside ourselves. They are created by fashion designers, advertising experts, priests, gurus, biomedical experts, and generals — mostly men. Except in the few cases where actual people, like Brooke Shields and Clark Gable, become the ideals, they exist only in a realm outside living bodies. Not possessing these ideals within ourselves, we cannot know what is best for us. We must rely on the supposedly expert knowledge of those who understand the ideals and how to apply them.

Each of us is influenced by a unique combination of ideals. The ones we choose depend on such factors as our gender, economic and social status, profession, and psychological and spiritual history.

Pause

When you are reading magazines or watching television, which bodies appeal to you? How do you feel about yourself in comparison to these bodies?

Standing in front of a mirror or looking at a photograph of yourself, ask these questions: Should my shape be different — my shoulders broader, hips narrower, buttocks or chest more or less pronounced? Is my spine the right shape? My pelvis and shoudlers at the correct angle? Am I too fat, thin, tall, or short? By comparison to what norms, or by what authority, have I come to think my body should be those ways?

TYPES OF IDEALS

Each religious tradition has a somatic ideal that reflects its particular theology. In some traditions, such as Hinduism, the ideal and its theological significance are explicit; in others, it must be inferred from art, ritual, and prescriptions for prayer.

Hatha yoga is easily the most ancient and sophisticated system for shaping bodies. Over millennia yogis developed hundreds of precisely defined postures that enable an aspirant to identify with every level of creation, ranging from plants and insects to divine avatars. Each asana, or posture, is designed to purify specific parts of the body and produce certain kinds of consciousness.

For example, *Kūrmāsana* is named after Vishnu's incarnation as a tortoise. It requires the yogi to draw his limbs in on all sides as a metaphor for withdrawing the senses from the world to calm the mind.

Figure 5

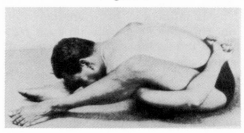

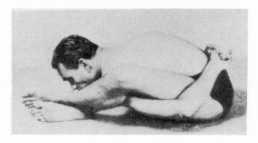

Practicing it is said to free the mind from anxiety amid pains, to make it indifferent to pleasures and release it from the grip of fear and anger. On the physical level, the posture is supposed to tone the spine, activate the abdominal organs, and soothe the nerves of the brain.[1]

The *vṛschikāsana* imitates the scorpion. It expands the lungs, stretches the abdominal muscles, and vigorously tones the spine. One master describes its psychological significance:

Figure 6

The head which is the seat of knowledge and power is also the seat of pride, anger, hatred, jealousy, intolerance and malice. These emotions are more deadly than the poison which the scorpion carries in its sting. The yogi, by stamping on his head with his feet, attempts to eradicate these self-destroying emotions and passions. By kicking his head he seeks to develop humility, calmness and tolerance and thus to be free of ego.[2]

The goal of the unified system of asanas practiced over a lifetime is to shape a person's restless bodily and mental impulses into the service of the quest for the Universal Spirit. The Upanishads promise that when the yogi has full power over his body, he will then obtain a new body of spiritual fire transcending illness, old age, and death.[3] The body is the source of confusion, pride, and the illusion that we are separate egos. In taming the power of the body, yoga frees the soul for the experience of reality.[4]

Buddhism radically simplified yogic practices, reducing its myriad designs to the image of Siddhartha becoming enlightened while sitting under the bodhi tree. That "sitting" constitutes one of the most powerful somatic ideals.

Figure 7

Buddha in Samadhi. Stone sculpture, Ceylon. 2nd century, A.D.

Even the simplest forms of Buddhism, such as Vipassana, teach that the absolutely quiet sitting body best serves the interests of meditation. It is difficult to grasp the idealistic character of this message because instructing students to sit in whatever way they like so that they can meditate for a prolonged period seems so innocuous. But the implication is that sitting quietly, ignoring the pains and impulses of the body, is the key to Truth. The authority in this tradition is not the body (conceived as *corpus*) but the mind; not "my" mind, but Mind as defined within a tradition of spiritual teachers in southeastern Asia.

When Buddhism migrated to Japan, its body image took on a more idealistic form. Zazen prescribed a specific place for pelvis, lumbar spine, head, and arms. The somatic form is thought to reflect the nondualistic character of Zen Buddhism. Suzuki Roshi writes: "These forms are not the means of obtaining the right state of mind. To take this posture is itself to have the right state of mind."[5] If the head is not in the proper place, one will not have the proper kind of awareness.

Figure 8

It is easier to understand the idealistic character of the apparently simple prescription for sitting if you consider a different somatic tradition. Sufism, for example, teaches that the vertical body spinning counterclockwise with an erect spine, and the arms in specific direction, in harmony with the cosmic rhythms of atoms and planets, is the source of the highest forms of consciousness.

Figure 9

The structure of hips, knees, and ankles necessary for prolonged sitting in the lotus posture is markedly different from that evoked by whirling.

Although Christianity does not have the elaborate descriptions of body structure found in the East, we can glean certain characteristics from church art, ritual, and spiritual practices. Christianity places more emphasis on the

Figure 10

vertical body, suggesting a reaching from earth toward heaven — the sentiment expressed in the lines of Gothic art and architecture.

Kneeling erect, prostrating, standing, and sitting in angular wooden benches characterize Western spiritual techniques. The line from foot through knee and hip to spine is straighter than one finds in bodies shaped for the lotus posture. The design of chapter rooms that has prevailed in Western monasteries during their eighteen-hundred-year history allows the monk or nun to kneel, sit, or stand for hours only within a narrow space defined by verticals and horizontals.

Figure 11

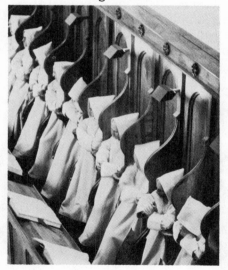

The Protestant Reformation was a somatic revolution. Prostrating and prolonged kneeling were eliminated as forms of worship. The ideals were demythologized, and rigid verticality became associated with good breeding and health rather than with spiritual perfection.

Religious ideals have often been connected with military

and therapeutic models. Yoga, tai chi chuan, and Zen were martial and therapeutic as well as spiritual arts, and the training of European soldiers was once modeled on monastic discipline. The Western military model has remained fairly constant since the fourth century B.C. It is a body designed for long marches in close-order drill, with a musculature trained to sustain and deliver heavy blows. But this design has more than functional value. Sucking in the belly, throwing back the shoulders, and stiffening the overall alignment constitute an effective model for inhibiting sensual impulses. A body thus shaped over years of training becomes an effective tool of national policies, unlikely to resist commands at inappropriate moments. The military ideal shares with the Western monastic ideal the goal of leeching out any individualistic impulses in the service of common action. Like its religious counterparts, it is a vertical, "straight" body from foot and hip through spine.

Figure 12

Figure 13

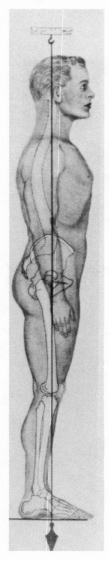

Since Western orthopedists have been concerned primarily with extreme cases of skeletal deformity, they have given virtually no explicit attention to what an "ideal" spine would look like. But one can tell from designs of so-called normal bodies and from orthopedic devices that medicine has borrowed uncritically from popular military notions of "straightness." This photograph from a widely used anatomical chart shows how the "normal" body approximates good military posture.

In Figure 14 orthopedics reveals its procrustean tendencies in the oubliettelike mechanisms it has devised to straighten patients. These are examples of nineteenth-century contraptions.

Figure 14

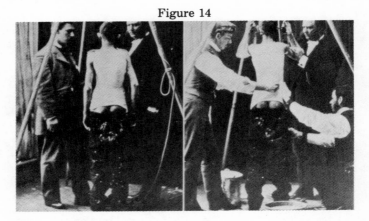

Plaster application by Sayre for scoliosis correction. (From Sayre Textbook on Scoliosis — 1870.)

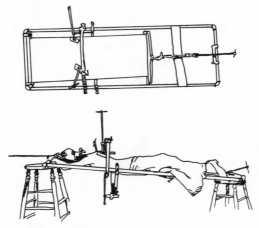

From Brackett and Bradford — 1890.

Figure 15

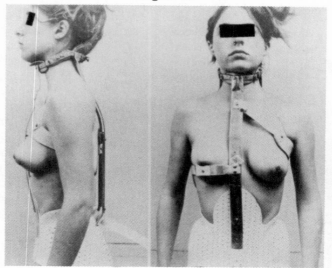

The modern Milwaukee brace with plastic pelvic girdle.

A textbook of orthopedics praises the commonly used Milwaukee brace, illustrated in Figure 15 as a landmark of progress in correcting spinal curvatures. I have worked with several adolescents whose pain and rigidity testified to the dubious value of such devices.

During the nineteenth and twentieth centuries, several people in Europe and the United States began to examine the importance of skeletal structure more carefully. Each of these researchers founded a therapeutic school based on an ideal notion of the body. Andrew Still and Daniel Palmer founded osteopathy and chiropractic, with its ideal spine.

Figure 16

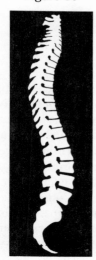

Figure 17

The Australian F. Matthias Alexander created a method for bringing people's spines toward this shape.

Figure 18

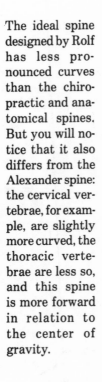

The ideal spine designed by Rolf has less pronounced curves than the chiropractic and anatomical spines. But you will notice that it also differs from the Alexander spine: the cervical vertebrae, for example, are slightly more curved, the thoracic vertebrae are less so, and this spine is more forward in relation to the center of gravity.

You may appreciate the significance of these therapeutic ideals for the day-to-day shaping of our bodies if you recognize that they are used in the design of school desks, automobile and airline seats, office furniture and equipment. They determine the relative positions of various components in the assembly lines of factories. We spend a good part of our lives unconsciously conforming to such shapes.

Consider, for example, this design for a computer station developed by Henry Dreyfus, the leading authority in what is called "ergonomic design," a method of designing the work environment to support what is thought to be the ideal posture for the bodies of the workers:

Figure 19

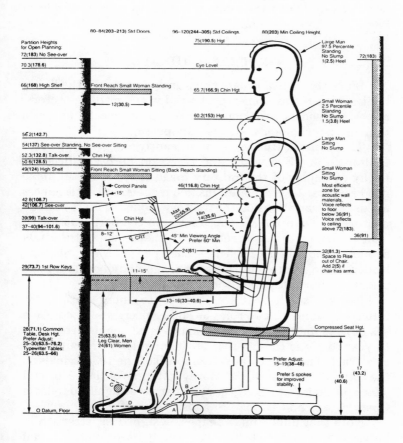

If you put yourself in this position and imagine adopting it for 40 hours every week, I think you can get a sense of the stillness it evokes in the pelvis, abdomen, and thorax. It also encourages a person to strain forward with their eyes and neck. No surprise that so many office workers complain of back pain and eye strain!

AESTHETIC IDEALS

Since their outlines are constantly before us, the living bodies of men and women portrayed in the media — actors, actresses, models, and sports stars — communicate the ideals that are psychologically most compelling.

The male aesthetic is significantly constant. Robert Redford, Clint Eastwood, and Harrison Ford are different fleshy examples of the military ideal; they would all make good soldiers. Male athletes, fashion models, and entertainers are usually variations on the military model, in more relaxed postures and clothing. Being rooted in the perpetuation of male power, institutions conspire to shape men to the kinds of molds thought to incarnate that power. Male religious leaders have created patterns for ritual body use, and male scientists have designed the medical ideals.

Female aesthetics are subject to more variability, reflecting women's two roles as child-raisers and decorative sex objects for men. Women's shapes must change from year to year to pique the interest of the men they are supposed to attract. The gaunt Carole Lombard of the 1930s yields to the zoftig Jane Russell and Marilyn Monroe of the 1940s and '50s, and that shape has gradually narrowed again to today's Cheryl Tiegs and Brooke Shields.

Note that unlike male shapes, designed for strength in battle and work, female ideals are rarely based on functional considerations. In a recent article in *The Journal of American Podiatry*, for example, a physician argues that women should be encouraged to use high heels even though they throw the body out of kilter, causing curvature of the spine and tensions in the knee and hips. These problems, he says, are a small price to pay for the sensuous effect of altering the whole anatomy, particularly the breasts and the buttocks: "For a male, a natural voyeur, the sight of a trim woman in high heels is an intravenous shot in his libido."[6]

IDEALISM AND ALIENATION

An ideal, like any technique for realizing it, is not necessarily a bad thing. Ideals have always had a valuable function in expressing our sense that we use only the smallest part of our potential, and body ideals inspire people to try new ranges of movement and muscular development. Western gymnastics, hatha yoga, and tai chi chuan are idealistic systems that have contributed to the well-being of generations of practitioners, improving both their muscle tone and their consciousness. By manipulating the spine in specific directions, chiropractic and osteopathy have alleviated ills that medicine could not cure. The Alexander technique and Rolfing achieve remarkable results by directing the spine according to different templates.

Of special value in this age of social fragmentation is an ideal's capacity to create social cohesion among its adherents. The design creates a commonly shared aesthetic, discipline, and language, giving a particular group a sense of shared goals. Those who espouse the Zen model, for example, share a common vocabulary and set of somatic techniques. Dedication to the bioenergetic model and its derivative therapeutic language gives its students a sense of being different from both militarists and students of chiropractic.

But these valuable experimental programs for working with our bodies and shaping our communities regularly get digested within the technology of alienation, to the point where one ideal comes to represent Truth, the shape that is thought to embody the truly human. Other ideals are said to be more primitive, aberrant, wrong, or even bad. Adherence to the ideal justifies the domination of one type of therapy over another, one form of spiritual practice, and in extreme cases, one kind of human being.

A person can become distracted by the competition among various ideals and miss their underlying unity in serving an authoritarian approach to bodily life. Being taught to conform our bodies to ideal shapes is a basic element in learning how to become good citizens. It doesn't matter what the ideal is. What is significant is the process of applying it.

Idealism derives from the Greek *idein*, which originally meant "to see." It is a way of thinking about reality based on a visual distance between the perceiver and the perceived. But this approach is visual in a very restricted sense. Unlike the vision of the artist, whose eyes wrap around their objects, feeling their contours and being shaped by the object, the idealist's vision takes place across a gap the observer creates between himself and the observed. The observed stands still, perhaps in front of a grid, while the physician, chiropractor, or posture expert stands at a distance, analyzing the "deviations" in the observed. "Experts" maintain that "professional distance" in order to emphasize the difference between those who know and the supposedly ignorant.

This alienated approach to the study of human beings typifies the methods used by the leading scientists of body structure. For example, the German Ernst Krestchmer, writing in 1921 about the relation between different body types and corresponding character types, described how he gathered his data. To make a list of 800 (!) specific variables: "We noticed, and immediately filled in point for point, the foregoing list, the patient standing before us in broad daylight . . ."[7] William Sheldon, the American pioneer in body typology, gathered his data on the structures of white males by having them stand rigidly erect with their feet together and their arms held straight. (Figure 20.)

In many school districts nurses routinely examine children for scoliosis (curvature of the spine). The child stands in front of the nurse, who sometimes asks the child to bend forward while he or she takes notes on the visible deviations from the ideally vertical line. Children judged to have scoliosis are given various prescriptions. In the most extreme cases, medical personnel advise parents to have metal rods

Figure 20

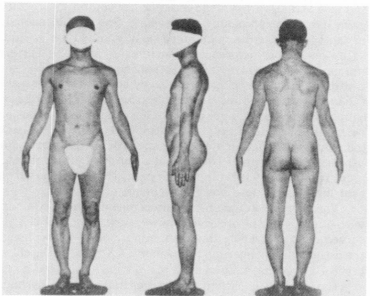

"Little tomcats. Lithe, compact, self-sufficient predators who can make their own way and can kill their own meat with help from nobody, and can do it even in the teeming cities."[8]

surgically implanted in the grooves on either side of the child's spine, whereas they suggest rigid corsets and exercise programs for less serious cases.

Like Kretschmer and Sheldon, these nurses are trained to stand at an impersonal distance and observe *the* body (not Susie's or Johnnie's body) as a collection of inanimate parts. They see this particular body only as something that is out of line in varying degrees with an abstract image, whose validity has been decided upon by some unknown "expert." They do not grasp the organic unity of the child by inquiring into how the scoliosis fits into his or her life or how it relates to other parts of the body and to the child's personal history.

Let us use Sister Ruth's case for an example. After her first back surgery, Sister Ruth began to receive significantly more attention from the nuns in her community, which gave her a great amount of pleasure. She simultaneously began to drink and smoke and nearly doubled her weight. "How else could I express my loneliness?" she says. During the next twenty years and three subsequent surgeries, she continued to enjoy being cared for by the community while the back pain persisted. She reports that it was easier to accept affection given in response to her illness than affection based on forbidden sexual attraction. To consider her back pain only as a function of the alignment of five lumbar vertebrae and the sacrum is to abstract a barely significant fragment from the full meaning of her pain. Treatment based on that diminished data had little chance of success.

The visual medium of idealism is meant to convey a nonverbal message: the observer knows something that the observed does not. The nurse examining a child for scoliosis, the orthopedist pondering over Johnny's X-rays, or the chiropractor discerning subluxations in a person's spine has a perspective that their patients can never have, except perhaps when the "expert" explains their "aberrations" in a photograph, video playback, or mirror. The "experts" have the necessary external standpoint and knowledge of the ideal in light of which judgments are made. They are thought to be free of the subjective biases that they say cloud our self-perceptions.

Idealism also contributes to authoritarianism by evoking a muscular rigidity which dulls our perception. Both the Greco-Roman and the Oriental ideal glorify stillness as the rational adult's conquest over the unceasing movement of the infant. Contemporary body idealists have based their designs on how the body should be organized if it is not to topple over while standing still, so it can maintain the status quo.

Ideal bodies are like algebraic equations, frozen in time, devoid of living flesh. They teach us to move through the world mechanically. A philosopher of art writes: "The more quiet the posture of the body is, the more apt it becomes to express the true character of the soul. In all postures which deviate too much from the state of rest, the soul is not in its proper, but in a forcible and forced state."[9]

One of the ways in which you can recognize that a particular design has passed from the realm of utility into the arena of power is the structure of idealistic language. Somatic idealists regularly speak of *the* body, rather than *my* body or *your* body. They see individual bodies as imperfect instances of the ideal pattern, which is the true body, and individual differences, the stuff of genius, as things to be corrected rather than qualities to be treasured.

Françoise Mezières, for example, the creator of a form of physical therapy popular in France, teaches that the ideal shape of the human body is found in Greek art of the classical period. She teaches her students "not to accept any treatment that is not directed toward that perfect form. The Greek artist didn't attempt to express psychological, mystical or political contradictions — but rather a corporal and moral unity toward which each mortal, out of self-respect, should direct himself. Any deviation from this description individuates a corporal deformity."[10]

Words like *deformity, ugliness, misalignment, aberration,* and so on are rooted in somatic idealism. These words are reserved for human bodies. One does not say that the Colorado River has an aberration where it has created the Grand Canyon or that a gnarled cypress growing on a windswept cliff is misaligned. The most sinister use of somatic ideals calls into question the humanity of those whose shapes are far removed from them — the palsied, paraplegics, dwarfs, Down's syndrome children, and others.

The founder of a contemporary school of body movement in California has designed a model of the body that is specific to the point of dictating an ideal location in space for each part of the body for each movement. One day I described to him a young girl with whom I have worked over a period of years. She has cerebral palsy. When she first came to me when she was seven years old, she had great difficulty in doing the most ordinary things: writing, walking, playing simple games. Her teachers complained about her performace in school. During the first session that I worked with her I had to restrain my tears because I was so moved by what I sensed was an unusually brilliant and loving person restricted by a difficult personal and family history. After four years of working with me and a number of other therapists, this girl functions at least as well as her peers: she skis, skateboards, plays the piano, and does well academically. But, I said to the founder of this school, she doesn't look anything like your ideal, nor will she ever. He replied, with a poignant sigh, "We must realize that certain people are not fully human."

Ida Rolf said of the significance of her template:

> The vertical line of Structural Integration is an abstraction, but it is more than an abstract vertical line. Probably you have never considered the possible significance of a *biological* vertical line. Have you ever realized that man is the only animal whose face is invisible as you look down directly on the top of his head? That statement is very significant. It is implying that the progression from ancient to modern man has been a verticalization which creates new relations within the creature.[11]

An obvious problem with this argument is that when you look down on the top of the head of a black person or a Latin American Indian, you can see a good deal of the face; you see a little less with most Western Europeans, and the least with East Indians and Chinese. Rolf was unaware that her reasoning was consistent with a long history of pseudoscientific argumentation supporting racism. At least as early as the eighteenth century anthropologists were claiming that the classical head had a vertical plane; the Negro head was tilted backwards and thus associated with smaller cranial capacity and less intelligence.[12]

None of these somatic ideals has more than an imaginary relation to our sweaty feet and sagging spines. Their "truth" is at best pragmatic; sometimes some of the ideals help us achieve health or increased consciousness. But one often hears the argument that a particular ideal is true in the sense that deviation from it, say in extreme scoliosis, causes pain and reduces vital capacity. Neck, shoulder, and lower-back pain sometimes accompany severe curvatures in those areas and is relieved by braces or surgery that corrects the deviations. But the chains of causality are unclear.

In many of my clients the major causes of physical pain have been stereotypical use of their bodies and repeated mechanical performance of particular actions, without sensitivity or interest. When a person gets bored with his or her work, when it becomes routine and creativity is dried up, he or she often starts feeling pain. The result is that the various asymmetrical curves in that individual's body become more pronounced. In adolescents I've worked with scoliosis often seems to be a consequence of growing emotional rigidity. As they enter puberty, their fears and awkwardness intensify the differences in their muscle groups, and their muscles take on rigidities which in turn compress the spine, exaggerating its curves.

Moreover, there is no patterned relation between asymmetry and pain. On one hand I have worked with people who are as close to perfect symmetry and alignment as anyone I have ever seen but who still experience intense, debilitating back pain. On the other hand are people like Isaac Stern, who has spent his life playing the violin, a highly asymmetrical use of the body. His body is extremely curved, yet he radiates a grace and bodily pleasure which most of us can envy.

The emphasis on perfect symmetry in the various ideals does not correspond with the design of the body itself. We have three lobes of the lungs on the right side of the chest, two on the left. The organs are asymmetrically arranged, the imbalance between the large masses of the heart and liver being particularly significant. The functions of the brain are asymmetrical. Most manual occupations and most sports require us to use our bodies in asymmetrical ways. Driving, writing, eating, and sitting comfortably are asymmetrical activities. The designers of somatic ideals, however, would have us try to construct a perfectly symmetrical shell of muscle and bone around this core of organic and neurological asymmetry.

In the previous chapter I mentioned the case of two-year-old Johnny, whose orthopedist prescribed a brace for the child's legs because of his assumption that he knew the ideal place for Johnny's knees and ankles. But if a person were to respect the uniqueness of Johnny's structure and his capacity for self-regulation, he or she would accept that Johnny is the one destined to find his proper stance in the world. That stance will be in a state of constant flux as he learns to respond with his genius to the requirements of an ever-changing milieu.

Infants are engaged in learning how to organize themselves to crawl, grasp, stand, and walk — processes requiring months of experimentation, feedback, and self-correction. All the while their bones and muscles are taking new shapes, depending on the direction of the experiments. If infants are given support and assistance in pursuing their often clumsy, usually brilliant explorations, they generally succeed in developing stable body structures. They require affectionate and supportive touching, and encouragement to continue experimenting even when they might be venturing near dangerous limits.

Several factors can impede this process. Forcing braces on Johnny would have kept him from learning how to develop his own support system. Sister Ruth's process was severely interrupted by orthopedic surgery and pain-killing drugs. She was not treated as a person whose bodily pains gave meaningful signs about conflicts in her life, but as a corporeal machine whose parts needed adjustment.

The constant impact of ideals, assisted by the design of our environment, school discipline, and regimentation of the workplace, gradually reduce our native sensual curiosity. Losing flexibility, we perceive less. Disconnected from awareness of our perceptions, we become more dependent on those who are supposed to know. From our earliest years we are taught that *they* know more than we, so we had best be still and docile children/students/patients/citizens, allowing the authorities to run the show, whether it happens to be a hysterectomy or a nuclear war.

CHAPTER SIX

DISEMBODYING THE ENEMY

Jonathan Schell writes of the extent to which our social policies have been cut adrift from their moorings in the biological:

> Four and a half billion years ago, the earth was formed. Perhaps a half billion years after that, life arose on the planet. For the next four billion years, life became steadily more complex, more varied, and more ingenious, until, around a million years ago, it produced mankind — the most complex and ingenious species of them all. Only six or seven thousand years ago — a period that is to the history of the earth as less than a minute is to a year — civilization emerged, enabling us to build up a human world, and to add to the marvels of evolution marvels of our own: marvels of art, of science, of social organization, of spiritual attainment. But, as we build higher and higher, the evolutionary foundation beneath our feet became more and more shaky, and now, in spite of all we have learned and achieved — or, rather, because of it — we hold this entire terrestrial creation hostage to nuclear destruction, threatening to hurl it back into the inanimate darkness from which it came.[1]

The most disastrous result of splitting mind from body and intelligence from perception, and of giving value to the former over the latter, is the topsy-turvy system of social values found in the recent history of human slaughter, which has been carried out by male "experts," justified by scientific rationalism, and supported by masses of citizens who have been trained to perceive only in the most truncated fashion. I selected the cases of Walter, the Los Alamos physicist, and Charles, the former CIA agent, for inclusion in this book because these men are typical of those who have produced our weapons and formulated policies for their use. They are not psychotics like Dr. Strangelove, nor warriors hardened in combat, inspired by macho ideals. They are reasonable, civilized men with wives and children.

In his book about our nuclear policy-makers, Robert Scheer writes:

> As I have come to know them I have been struck by this curious gap between the bloodiness of their rhetoric and the apparent absence on their part of any ability to visualize the psysical consequences of what they advocate . . . . . Discussions of global violence come to seem absurdly — not to say hideously — abstract when these theorists discuss the prospect of mass destruction as something apart from actual metal tearing human flesh and bodies radiated to oblivion as millions upon millions die either in a blinding flash or unimaginable prolonged agony.[2]

Scheer tells of an evening when Richard Perle, assistant secretary of defense, invited him to his home for a discussion over Scotch and sodas. They had a quiet and genteel discussion about nuclear strategies until Perle's two-year-old son climbed up onto his father's lap and smashed him in the face, causing Perle to spill his drink. Perle jumped up, enraged, and dumped the infant on the floor. Scheer comments that "a single episode of tolerable mayhem quite shattered the dispassionate, not to say anemic, civility of our consideration of untold millions who would have to endure far greater suffering than a small fist in the face."[3]

Walter, Charles, Richard Perle, and men like them represent the most refined products of the technology of alienation. Their coolness in the face of mass extermination testifies to the true nature of the pressures brought to bear on us from infancy through adulthood to shift our gaze outward and upward — outward from our own perceptions and feelings toward the authority of official experts, and upward from the palpable world of flesh and growing things to the realm of supposedly more noble abstract ideals.

In the *Bhagavad Gita*, the hero Arjuna, often held up as a prototype of the spiritual seeker, is supposed to enter into battle against his relatives, friends, and teachers to take some property won in a throw of dice. He complains that it would be better for him to become a beggar than to kill these revered people. The divine Krishna replies in words whose sentiments were later echoed by Plato, Saint Paul, and the strategists of the Kremlin and the Pentagon:

> It is said these are temporal bodies
> of an embodied one — eternal,
> Indestructible, beyond measure.
> Therefore fight, son of Bharata.
> Who thinks of him as slayer
> and who thinks that he is slain
> Do not rightly understand.
> He slays not, nor is slain.[4]

Arjuna keeps raising the obvious point: that it is repugnant to kill loved ones over a quarrel begun in a dice game. But Krishna spins a marvelous metaphysics to convince Arjuna that he is deceived because he attaches reality to the perceivable world. Arjuna finally capitulates, renouncing his senses in favor of that obscure metaphysics which argues that killing is not killing.

Arjuna's capitulation presages the history of humanity's giving up its ability to see. Many Germans, Poles, and French didn't really see their Jewish neighbors being deported to the concentration camps. Many Russians and Chinese were blind to the massacres of millions of their fellow citizens by Stalin and Mao. Many Americans looked at photographs of the victims of Hiroshima and Nagasaki without fully experiencing their meaning. All these people, of course, "saw" in the Cartesian, mechanistic sense of the word. Sensory data from the bodies of victims impinged on their eyes and were ineluctably transmitted to form images in the brain. But that kind of seeing is to human sight what the slight wiggle of one's little finger is to the full range of manual dexterity. The sensory impressions did not travel into these people's sinews and guts where they could evoke the courage to exclaim: No more! Stop this insanity! Unintegrated into the entire feeling body, the sensory data were not powerful enough to support an effective consensus to oppose policies that were rationally justified by "experts." Millions of otherwise decent citizens throughout the world renounced their basic humanity to support inconceivable atrocities. That vast moral failure was based on impaired vision.

A young army sergeant recently appeared on a CBS documentary devoted to analyzing U.S. defense policies. He was one of a three-member team at the bottom of a missile silo in North Dakota, charged with the job of actually firing an ICBM when ordered to. He seemed gentle and good-humored. Dan Rather asked him how he felt about his job. He straightened up, beaming: "I feel very proud to have such a key place in our country's defense." But when Rather asked him where his missile was aimed, his eyes lowered and his shoulders slumped. "I don't know where it's aimed," he replied with a note of sadness in his voice, "and I don't want to. I might find out that someone like me is on the other end and I wouldn't want to fire."

I am talking here about the phenomenon of denial, the perennial tendency of human beings to pretend that what lies in front of them does not exist. Several people, from Freud to Robert Jay Lifton, have analyzed the psychological aspects of denial, the ways in which we construct screens to shut out memories of unpleasant personal and collective events. In this chapter I am interested in the sensual basis for that denial as it relates to overlooking mass destruction.

Pause

What images come to mind when you think of trying to explain nuclear war to a seven-year-old child who asks you about it?

Recall photographs you've seen of dead bodies at the German concentration camps or at Hiroshima and Nagasaki. What feelings do those images evoke?

Are there devices you use to reduce the impact of these images?

When you read or see television analyses of the effects of a nuclear attack, what happens in your body?

Are there voices within you that say it is hysterical and irrational to dwell on such images? Where do these voices derive their authority?

Freud understood the connection between the willingness to kill and immediacy of perception. In his "Reflections on War and Death," he argues that there seems to be an innate urge to kill which remains at the subconscious level and usually surfaces only in fantasy. But there are conditions under which the most decent citizen would be willing to kill. He refers to Rousseau's musing about whether he would kill an old mandarin in Peking if he could do it by a mere act of will, at great profit to himself, and remain undetected. Rousseau decided he would do it without hesitation.[5] Freud's view of humanity, which may seem overly cynical, has been sadly substantiated by fifty years of the evolution of warfare since he wrote that essay.

For the first time in history, Rousseau's kind of killing has become a reality. Present weapons technology does not require us to see the enemy. Accustomed to our own disembodiment, we don't even have to think of the enemy as embodied. Our weapons are capable of destroying abstract populations in which flesh and blood are not reckoned. Removing defense strategies from their anchor in a bodily universe has unleashed a killer instinct that might have surprised even Freud.

Before the 1940s, war remained embedded in the psysi-
cal. It was impossible to indulge that secret desire to kill by
a mere act of will without being detected. One knew one was
killing. Learning to be a warrior meant gaining an intimate
sense of one's own body and a heightened ability to perceive
the bodies of the enemy. Training for war required a soldier
to maximize his muscular strength, capacities for endurance,
and abilities to see and hear, because the warrior had to be
keenly aware of the location and movement of his enemy.
Warriors had an immediate perception of the connection
between their acts and the effects of those actions on their
victims. They witnessed the deaths of their comrades as well
as of their enemies. The immediate contact between the
bodies of warriors and victims made killing an emotionally
weighty affair. To be a warrior legitimized by the community meant that one had to pass through rigorous physical,
moral, and spiritual initiations involving costumes, rituals,
rewards, community festivals, and religious services. Passing
through that training was a symbol of becoming a man.
"Dulce et decor est pro patria mori," wrote Horace — "It
is sweet and fitting to die for one's fatherland."

With rare exceptions, the male heroes of mythology are
warriors. The *Iliad* and the *Odyssey*, the Arthurian legends,
Beowulf, and the *Chansons de Roland* are epics about war
as the matrix for manly virtue. The very word *virtue* comes
from a word that means both manliness and strength, and
is associated with courage in battle. The heroes of modern
myths too are warriors, from *A Farewell to Arms* to *Star
Wars*.

The model for the spiritual seeker has often been the
courageous warrior. Saint John describes Jesus as the cosmic
killer:

> Justice is his standard in passing judgment and in
> waging war. His eyes blazed like fire, and on his head
> were many diadems . . . He wore a cloak that had been
> dipped in blood, and his name was the Word of God.
> The armies of heaven were behind him riding white

horses and dressed in fine linen, pure and white. Out
of his mouth came a sharp sword for striking down the
nations. He will shepherd them with an iron rod; it is
he who will tread out in the winepress the blazing wrath
of God the Almighty . . . (*Rev* 19: 11f).[6]

Saint Paul compares the inner life of the Christian with
preparing for battle:

Put on the armor of God so that you may be able to
stand firm against the tactics of the devil. Our battle
is not against human forces but against the principalities and powers, the rulers of this world of darkness,
the evil spirits in the regions above. You must put on
the armor of God if you are to resist on the evil day;
do all that your duty requires, and hold your ground.
Stand fast, with the truth buckled around your waist,
and integrity for a breastplate, wearing for shoes the
eagerness to spread the gospel of peace and always
carrying the shield of faith so that you can use it to put
out the burning arrows of the evil one. Take the helmet
of salvation and the sword of the spirit, the Word of God
(*Eph* 6:11–17).[7]

The Christian mystical tradition portrays Christ and
Satan battling over possession of one's soul, and Christians
are encouraged to inflict pain on themselves to subdue the
enemy within, a psychodrama imitating the action of soldiers
in combat. Ignatius Loyola's visions, which inspired him to
create the Jesuits and to teach the Catholic leaders of Europe,
depicted Christ as a leader of armies marshaled against the
armies of Satan.

Similarly, Islamic theology was born of Mohammed's
slaughtering of his enemies. Its central image is the holy war,
a spiritual task of killing and maiming those who violate
Islamic law. The *Bhagavad Gita* also is cast in the context
of killing. The art and literature of Tibetan Buddhism are
filled with images of wrathful deities brandishing weapons
to destroy the purveyors of illusion while round-breasted

and -buttocked *shaktis* dangle from their erect penises.

This is not to say that past warfare was as glorious and heroic as our myths have made it out to be. The history of war is one of rape, pillage, the slaughter of innocent victims in the bloodiest ways, and brutal torture. But past generations had to live with their perceptions of killing, sometimes embedded in nightmarish guilt fantasies and psychoses. They had to go to great lengths to justify destruction. Killing was officially condoned only if it was directed against specific people, aggressors or supposed violators of the law. Women, children, and the infirm were not legitimate victims, and warriors who indulged in wanton killing and were found out were reprobated. Citizens from Tokyo to Paris to Civil War Atlanta had an immediate sense of what killing was about. If they had not directly witnessed people being killed in war, they saw war's results in maimed and wounded neighbors.

Whether the warrior is any longer a viable organizing paradigm for life, particularly for male excellence, is a complex question not directly related to the themes of this chapter. What is relevant is to note the shift away from the sensual character of warfare, a progressive disembodiment of those who are labeled the enemy. Dematerializing war has escalated the amount of wanton destruction a supposedly civilized population is willing to tolerate.

As I have noted, attitudes toward officially sanctioned mass killing began to shift during the 1940s. The distance between killers and their victims became so great that the fantasy of killing the old mandarin in Peking became a reality. Warfare was transferred from the heroic realm of morality and religion to the supposedly more rational world of science and business management. Kurt Vonnegut writes of this transition:

'Death Before Dishonor' was the motto of several military formations during the Civil War — on both sides. It may be the motto of the 82nd Airborne Division right now. A motto like that made a certain amount of sense, I suppose, when military death was what happened to the soldier on the right or the left of you — or in front of you — or in back of you. But military death now can easily mean the death of everything, including the blue-footed boobies of the Galapagos Islands.[8]

The shift began with the Holocaust. Ordinary German citizens, from those close to the centers of power to simple shopkeepers, succeeded in blinding themselves to the extermination of millions of other decent Germans. A survivor of Bergen-Belsen writes that on her march to the camp, when she and her fellow prisoners were marked with the yellow Star of David, people in the villages did not seem to see them. She describes the camp itself as having an air of extreme rationality with the gas chambers designed to make the victims largely invisible to others in the camp. The prisoners' terror was only infrequently witnessed by those who had logically and with great precision planned their death. "This 'making invisible' of other people," concludes Professor John-Steiner

is that lesson, or burden, which I see as our most significant heritage of the Holocaust. Because it is only by making one's enemies invisible that we can escalate our abilities as warring nations. It is precisely for that reason that we are the only species that can consider destroying millions of our fellow beings in overwhelmingly rational, planned and efficiently calculated ways.[9]

That same blindness to the obvious was present in American attitudes toward Hiroshima and Nagasaki. The Defense Department, in alliance with physicists who supposedly embodied the highest levels of rationality, carefully preserved the largely nonmilitary populations of those cities from conventional bombing for the sake of experimenting with their new bomb. The physicists from Los Alamos who went to research the aftermath of the atomic bombings reported that after two days they stopped feeling the horror of what they saw around them. On American radio, Fred Allen quipped that Hiroshima looked like Ebbetts Field after a Dodgers game. Consciousness of the annihilation of half a million noncombatants and the mutilation of hundreds of thousands of others was conveniently suppressed until nearly thirty years later, when medical and psychological studies of the victims began to appear in popular literature. The suppression was sensual as well as psychological and political, and was made possible by the training that teaches us to drug our senses.

The 1940s marked a scientific breakthrough in using large populations of human bodies for weapons experiments. In 1981 investigators discovered that during the forties the Japanese had experimentally killed about three thousand humans, including Americans, with biological weapons. The victims were subjected to huge doses of plague, anthrax, and smallpox germs. Some were killed by radiation poisoning; others were pumped full of horse blood and cut up while alive. The U.S. Defense Department granted immunity to the director of the experiments, General Ishii Shiro, in return for the results of the experiments, arguing that "since any war crimes trial would completely reveal such data to all nations, it is felt that such publicity must be avoided in the interests of defense and national security."[10]

We may think that this is an exception, or argue that since the experiments had already been done we should make use of them. But the implicit condoning of torture and killing for the sake of weapons research is in line with our own policies of experimentation with nuclear explosions. Hundreds of thousands of American soldiers, told by our government that they were being trained for combat, were used as guinea pigs in the Pacific and Nevada testing sites, and nuclear experiments were continued despite evidence of danger to inhabitants of cities near the testing sites and to the natives of the Pacific atolls. The government also had information that workers in the nuclear weapons laboratories at Hanford and Oak Ridge were being subjected to dangerous levels of radiation, but the information was suppressed for the sake of national security.

The attitudinal shift that began in those years capitalized on a perennial tendency to flee the reality of flesh and blood and enter the realm of abstraction. The new military technology removed its operators from any relation to perceptible human bodies. Long-range bombers, new howitzers, and missiles placed vast distances between combatants. Biological, chemical, and nuclear weapons are not designed to eliminate discernible individuals or groups but are effective at the level of numbers that can hardly be imagined: troop divisions, populations, nations, and even groups of nations.

One would expect that as people increasingly denied the reality of the body symptoms of that denial would appear at some level of national policy. They did. A surreal quality began to turn up in military discourse, where doubletalk, coded languages, and discourse that seemed far removed from reality became the rule.

I recall an incident in 1965. I watched Robert McNamara, then secretary of defense, being interviewed on television about whether we should spend the millions of dollars necessary to construct an antiballistic missile system. In between ads for deodorants and denture stabilizers, he sat calmly in his gray flannel suit arguing the pros and cons of the decision in terms of dollars and lives. If we failed to build the system, 40 million Americans could be killed in a Russian attack; if we spent $100 million on the system, *only* 20 million Americans would be killed, he said. He sounded as if he were discussing whether to open a new Ford assembly plant.

In Vietnam, bombings and shellings were often planned by computers and based on probabilities of troop locations rather than on forward sightings of troop movements. There were no clear boundaries between civilians and combatants. It was not a heroic war in which our soldiers were killing a clearly defined enemy for a lofty national purpose. There were no heroes to remember; the lackluster Westmoreland succeeded Patton, William Calley succeeded Audie Murphy, and "Apocalypse Now" succeeded D-Day.

Herbert Marcuse wrote of the national attitude produced during the Vietnam era:

...the presentation of killing, burning and poisoning and torture inflicted upon the victims of neocolonial slaughter is made in a common-sensible, factual, sometimes humorous style which integrates these horrors with the pranks of juvenile delinquents, football contests, accidents, stock market reports, and the weatherman. This is no longer the "classical" heroizing of killing in the national interest, but rather its reduction to the level of natural events and contingencies of daily life. The consequence is a "psychological habituation of war" which is administered to a people protected from the actuality of war, a people who, by virtue of this habituation, easily familiarizes itself with the "kill rate" as it is already familiar with other "rates" (such as those of business or traffic or unemployment). The people are conditioned to live with the hazards, the brutalities and the mounting casualties of the war in Viet Nam, just as one learns gradually to live with the everyday hazards and casualties of smoking, of smog, of traffic.[11]

That "psychological habituation of war" enabled the public to accept as reasonable a fantastical series of civil defense policies initiated during the 1950s. Speaking of nuclear weapons as simply larger versions of conventional weapons, the government created the bomb-shelter craze. Schoolchildren were taught that they could save themselves from injury by diving under their desks while their teachers pulled down the blinds. Large cities mapped out escape routes into the country for citizens to use if they discovered that their city was about to be bombed. Hospitals were required to submit plans for caring for the casualties of nuclear attack. The Postal Service developed plans for distributing postage-free change-of-address cards to displaced survivors. The Federal Reserve System began an elaborate series of mock tests and analyses of how to preserve "the financial basis of democracy" in the event of an attack; over the past twenty years they have been asking such questions as "How we will print money if all of our capacities for printing have been destroyed?" "How will mortgage payments, taxes, and interest collection be ensured?" "What kinds of tax structure can be implemented if W-2 forms cannot be used?"

The government constructed a special vault to preserve the Declaration of Independence and the Constitution as a witness for postnuclear mutants, and large underground complexes for housing the government and the Federal Reserve System in the event of attack. A special Boeing 747 is to carry away the President and the heads of other branches of the government. And a government supplier manufactured a "Nuclear Calculator Set," a small circular slide rule. When you are standing outside and happen to see a nuclear blast, you calmly estimate the size of the mushroom cloud and its angle in relation to you. The slide rule will tell you how much you're being radiated. If you happen to have been in a shelter for several months, you can use the other side of the slide rule to determine how much radiation the material of the shelter is permitting to enter at any specified length of time after the blast.

In spite of these "preparations," at least as early as 1964 the Rand Corporation had made public an analysis showing that a one-megaton warhead would virtually destroy the Los Angeles basin. The initial impact would destroy the center of the city some twenty minutes after the missile had been launched; then, within hours, shock and heat waves would destroy large segments of the population. People in underground shelters would be incinerated, suffocated, or radiated on emerging. In the following days, radiation, the poisoning of water supplies, and lack of food and medical help would account for the deterioration of the population to the point where there would be only a few unhappy survivors. The image of two million people driving out of Los Angeles, with a twenty-minute warning, to some remote area large enough to protect them is one of the more obvious symptoms of collective denial of reality.

What I would like to note here is that the denial is not simply "mental" or "psychological." It involves prohibiting the images of these possible disasters from entering our nervous and digestive systems so that we allow ourselves to feel the sensual experience of the reality we are allowing our government to construct.

This book was born out of my shock at discovering the depths of my own denial. It happened in 1980, when I was working in Paris and traveling throughout Europe. A well-known French therapist had invited me to work in her elegant Montparnasse studio, where she and several disciples practiced the Mezières Method, a manipulative therapy aimed at forcing the body into perfect alignment. Having read my book *The Protean Body,* she supposed I would be a useful ally in straightening people out.

Virtually every one of my clients in Paris bore the scars of war. One woman with arthritic knees, for example, had been unable to walk when she was liberated from Auschwitz as a young teenager. Some years later, in Tel Aviv, she met Moshe Feldenkrais, who taught her how to walk again. A fashion designer whose muscles were extraordinarily contracted spoke bitterly of the night when he had been fourteen years old and the concierge had brought the Gestapo to his family's flat; his parents had been taken away, never to be seen again. Another woman, tense and cynical, told me of how her aristocratic family had been part of a network that harbored Jews on their escape routes out of Europe. When she had been four years old the Gestapo had arrived in the middle of the night and removed her parents, whom she never heard of again. She had been left alone in their chateau with her sister and servants for six months before other members of her family found them. I worked with a duke whose body was filled with shrapnel from French colonial wars in Indochina and Algeria. Another of my clients was a gentle Egyptian economist whose body was covered with scars from his days as a political prisoner under Nasser.

It began to disturb me that these and others were not coming to me to deal with their bitterness and despair; they were asking me to straighten their backs and make their shoulders more symmetrical. At the end of a day of touching these scarred people, I was often on the verge of tears. Riding home on the Metro I read of new outbreaks of French anti-Semitism: Jewish youths had been beaten, Jewish shops had been destroyed. In the evenings I watched French television's numerous documentaries on such events as Pol Pot's attempts to exterminate the Cambodian people, the massacre of two million Nigerians, and Khomeini's incitement of a new holy war against the enemies of the Shi'ite Allah. At home Ronald Reagan was campaigning for President by appealing to American fears that we were becoming too weak and needed more nuclear weapons. Russian suppression of dissidents had become more virulent, and the Soviets had invaded Afghanistan and were putting pressure on Poland to restrict the labor movement.

On periodic holidays my wife, my stepdaughter, and I took trips to different parts of Europe. Everywhere I saw firsthand what I had previously known of only through books: lands marked by the passage of armies — Julius Caesar's, Charlemagne's, Napoleon's, Patton's. It was often easy to see how modern buildings had been constructed on the rubble of previous wars. Every village had its monument listing its war dead. We visited the old Jewish quarters in places such as Venice and wept in front of plaques listing the schoolchildren who had been taken away to the camps.

I got a feeling for history that was unavailable on the Pacific Coast, whose only ravagers had been handfuls of Spanish soldiers, housing developers, and oil drillers. I was seeing and touching what it meant to give precedence to abstract ideologies rather than human flesh and blood. Within this barrage of perceptions about war I began to glimpse the relationships among ideal bodies, passivity, authoritarianism, and perverted social values. Pushing people toward ideal forms seemed to be one more device for convincing people that they don't know what is best for themselves.

These ideas took clearer shape when I returned to this country, to find a new President who went so far as to extol, in front of thousands of cheering young Christian zealots, a father who said he would rather have his infant daughter die believing in God than live as a Communist.[12] The new secretary of defense, formerly a quiet-spoken radio-station executive, had created a philosophy of "integrated warfare," articulated in the "Fiscal Year 1984–1988 Defense Guidance." According to this perversely holistic concept, the United States must be prepared to battle the Soviet Union simultaneously in Western Europe, on the Japanese and South Korean fronts, and in the Persian Gulf. The Navy must be prepared to fight simultaneously on all the major oceans. In any of these battle zones we must be prepared to use an integrated combination of nuclear, biochemical, and conventional weapons. The estimated bill for this kind of defense is $2 trillion over five years, or $8700 for every American citizen.

Cap "The Knife" Weinberger justified this new military holism and its lavish expense with this argument: "America is an island nation, with long and vital lines of supply and communication to Europe, Africa, Asia, and Latin America for crucial trade and essential strategic materials."[13]

Imagine a young man with long hair and a beard, dressed in Levis, saying that we need more defense because living in America is like living in Tahiti or the Orkney Islands. We have few resources of our own; we rely on long and tenuous lines of trade with the larger continents. Such a man would be judged a schizophrenic; people would say his brains had been fried by drugs. But the gray business suit, the tenor of seriousness, the official title, and the support of military experts with secret information give Weinberger's odd philosophy the aura of rationality.

Replying to a television reporter who commented that the casualties in an "integrated" battle would be unbelievably high, an American general nodded with a poignant smile. "That's the trouble with the integrated concept; we have to have several backup divisions to replace the ones that are destroyed."

The quintessence of the holistic defense concept is in space weaponry. The Air Force, which has been lobbying to change its name to the Aerospace Force, will spend $5.5 billion in fiscal 1983 on developing this wing of national defense. That figure does not include NASA expenditures on the space shuttle, nor the budgets that the Army and Navy allocate to space weapons.[14] "Directed energy weapons" being designed for placement on satellites circling the earth will be capable of virtually instantaneous obliteration of targets on earth. But "target" in this technology bears no discernible reference to human bodies. Peking, rather than Rousseau's mandarin, can be eliminated without trace. We have nearly succeeded in objectifying our worst fantasies about God: an observer who is always "out there," prepared at any moment to send us into an eternal holocaust.

In his televised address to the nation on March 23, 1983, President Reagan pleaded for support for his defense budget. With tears in his eyes he argued that the new space technology "holds the promise of changing the course of human history, offering a new hope for our children in the twenty-first century." Reagan's image of this peaceful world is one in which those lucky children will be protected from the Communist threat by living under a sky net of lasers, microwave systems, and projectiles.

The surrealism of defense strategies spills over into negotiations for arms reduction, another symptom of denial. Successive administrations have said that we need to increase our nuclear arsenal even though we are now capable of destroying the Soviet Union several times over, even if it attacks first. The argument is that we must have significantly more weapons than the Russians before we can effectively negotiate to reduce our weapons. Herman Kahn, a principal architect of present defense policies, believes that we must have a "not incredible," in contrast to a "credible," nuclear first-strike capability. There is a distinct difference between the two, he asserts. "You really can't achieve a capability which looks like it would be used, but you can achieve a capability which the other side cannot feel will not be used if he's too provocative." Kahn tries to convince us that he's making sense by asserting, "And the term 'not incredible' really carries an extraordinary amount of weight."[15] One congresswoman says this is "like two boys sitting in a large pool of gasoline, one with seven matches, the other with five."

The Defense Department recently announced plans to manufacture thousands of tons of a new, "safe" chemical weapon that is so poisonous that enough to cover a small freckle is fatal. In a letter to Congress, President Reagan termed the action "modernization of our retaliatory capacity which will provide strong leverage toward negotiating a verifiable agreement banning chemical weapons."[16]

Holistic warfare is not run by samurai, knights of the Round Table, or brave soldiers hardened on the battlefields. The new warlords alternate among the board rooms of the international corporations, offices of the Pentagon, and government laboratories. They destroy without getting their hands dirty. Trained in law, management, or physics, they are cooly rational enough to be unmoved by emotional images of flayed and radiated bodies. They never display feeling, and would never weep over their wounded brothers like Achilles over Patroclus. They show no signs of the moral perplexity expressed by Arjuna when pleading with Lord Krishna. Their agents are no longer courageous soldiers capable of making long marches in sub-zero weather and digging into trenches for months at a time, but passive men who have the patience to sit for years at the bottom of a missile silo waiting for orders to push a button.

The president of the Old Crows, an elite organization of scientists, engineers, business executives, and military strategists, argues that future conflicts will be determined not by old-fashioned warriors but by espionage experts — James Bonds with computer-science backgrounds.[17] The Defense Department has begun to economize by training soldiers with computer-simulated battles rather than the more expensive field maneuvers. One defense contractor, for example, has developed the Lasertour: personnel learn how to operate tanks by sitting at control panels in front of a large video screen, which displays various possible battle scenarios. You can buy one for your child at Neiman-Marcus.[18] President Reagan has praised video games for teaching young boys the skills they need to become good soldiers.[19]

This tendency to make mass-killing an abstract, ethereal

affair is assisted by coded languages that mask the reality of their referents. A government consultant on negotiation strategies writes about how that language removes the President from the physical reality of what he is doing: "There is a young officer who follows the President with a black attaché case containing the codes needed to fire nuclear weapons. I envisioned the President at a staff meeting considering nuclear was as an abstract question. He might conclude, "On SIOP Plan One, the decision is affirmative. Communicate the Alpha line XYZ.' Such jargon holds the results of his action at a distance."

This same consultant has suggested a strategy for forcing the President to be more aware of the fleshly enormity of his actions.

> Put the code number in a little capsule and implant the capsule right next to the heart of a volunteer. The volunteer would have a big, heavy butcher knife to carry as he or she accompanied the President. If ever the President wanted to fire nuclear weapons to kill tens of millions of people, he would have to start by killing one human being, personally. The President would have to look at a human being and realize what death is — what an innocent death is. Blood on the White House carpet. Reality brought home. When I suggested this to friends in the Pentagon they said, "My God, that's terrible. Having to kill someone would distort his judgment. The President might never push the button."[20]

A former CIA analyst who was responsible for some of the most important evaluations of Soviet nuclear capacities and who had witnessed the atomic detonations at Bikini spoke to Robert Scheer of his son, who joined the Marine Corps. He told about his son's enthusiasm, his expert marksmanship, how he does hundreds of sit-ups a day and runs miles in a few minutes. He described how he tries to imagine what it would be like for his son in a nuclear war, according to the abstract calculations he has been doing.

He's up close — not like me, far away, protected
with a water wash-down system. He's right there, he's
on the front lines. And I'm saying to myself, he's in serious trouble. I can see a variety of things that are going
to happen to him, either quickly or afterwards, that are
not pleasant. And then I put myself back in this theoretical, strategic stuff, where these guys just calculate
megatonnage. But my son is fried.[21]

When people could see their victims, they knew that in
the course of winning the battle they could lose their souls.
Coded languages and computerized weapons cloud that
vision, allowing people to be unaware that they are engaging
in or preparing for the killing and maiming of unimaginably
large numbers of ordinary men, women, and children, for the
sake of goals that are rarely questioned. The enormous
slaughter that is now an integral component of the defense
strategies of the supposedly most civilized nations is an
extreme symptom of a tendency throughout history to hide
from ourselves the fact that we are embodied. Instead of
being treated as a cancerous aberration, that tendency has
been enshrined as rational and authoritative.

The cable-TV entrepreneur Ted Turner recently made a
plea that echoed in modern dialect the plea that Arjuna made
to Krishna the Destroyer at the beginning of the *Bhagavad
Gita:*

Why don't we grow up, sit down, and say the hell
with it? Just keep the weapons we've got. And it would
be great to keep them. They'll be tremendous museum
pieces at some point in time. We'd survive on the planet
and take that budget we're putting into that crap and
put it into cleaning up the ecology and start planting
trees and things along the highways, fruit trees so we
could feed people and so forth. Put it into solving air
pollution and water pollution and intelligently farming
the land and curing soil erosion problems and dealing
with the world population explosion and so forth.
Take those billions and then, you know, in a thou-

sand years from now people can go to the Smithsonian and see those missiles and warheads and get down on their knees and thank God that people weren't so stupid.

This could be the most beautiful place... we've got a beautiful thing here: elephants, ants, mice, rabbits, squirrels, bluebirds. Why don't we preserve them? Preserve our environment and stick around for a while. It's a pretty nice life, you know.[22]

That Krishna's and Weinberger's metaphysics hold sway over the sensual humanity of Arjuna and Ted Turner is a sign of how far we've been disconnected from our moorings in the sensible.

CHAPTER SEVEN

A WOMAN'S WAY

The mind–body fracture that prompts us to discredit our sensual wisdom is reflected in ancient divisions between men and women. Nowhere is the illogical quality of what passes for our social logic more obvious than in popular stereotypes about women. We have constructed a mythical world in which men in gray flannel suits who sit at computers or in board rooms planning mass slaughter are considered prototypes of reasonable behavior, capable of directing the destiny of our planet, while women who take care of their homes and children or who work as secretaries, or nurses, often alone in their old age, are thought to be irrational sources of error or even sin, unfit to make any major decisions about public affairs.

Recall Robert Scheer's picture of Richard Perle, sitting calmly in his study sipping Scotch and soda, speaking quietly about the extermination of millions of people, and flying into a rage when his baby climbed on his lap; or Robert McNamara, coolly figuring the odds of sacrificing tens of millions of people; or Caspar Weinberger's notion of holistic warfare; or Krishna's sermonizing that the noblest manly act is to kill one's relatives and friends in the name of duty. To be more melodramatic, we might add Mao, Hitler, Stalin, Khomeini, Jim Jones, and Charles Manson. Against the backdrop of these males, who are not anomalies in the history of power, think of the popular myths about women that have prevailed in every culture. Like the bodies that emerged from their *bottichs*, women have been consistently identified with the evils that are thought to come from submitting to natural impulses and emotions.

Eve and Pandora are prototypes, as is Circe, weaving her hypnotic spell over poor Odysseus. Men often picture women as seductive creatures whose open thighs exude sweet odors that lure ingenuous men into dark places that reason would counsel them to avoid. An Indian sage writes:

> It needs no great imagination to know of the mystery of women and the dynamic story of their amazing secret life. Women who hold high positions in society, appear refined, cultured, educated, virtuous and innocent by day, but represent a totally different picture by nightfall. By night, under "neurotic" compulsion they commit the immoral act, bringing with them human degradation, destruction and sometimes death. A woman is not drunk when she commits the act. She gets befuddled by the powerful "sex currents" and "sex waves" that are ever present in the atmosphere. This is the first indication of her desire — her sex desire — which presses upon her loins and sweeps upwards into her body, filling the mind with one impelling thought.[1]

Greek shepherds divide their universe into men and sheep in opposition to women and goats. Although they do not identify women simply as devils, they feel that the nature of women's sexuality, "which continually threatens the honour of men, makes them, willingly or unwillingly, agents of his [i.e., Satan's] will."[2] Octavio Paz contributes a Latin view: "Whether as prostitute, goddess or *grande dame* or mistress, woman transmits or preserves but does not believe in the values and energies entrusted to her by nature or society. She is a domesticated wild animal, lecherous and sinful from birth, who must be subdued with a stick and guided by the reins of religion."[3]

Too close to the body, women are unclean. They bleed,

their breasts swell, they ooze fluids and smells. They must be kept from the centers of ritual purity. "When a woman has her menstrual flow," writes the author of Leviticus, "she shall be in a state of impurity for seven days. Anyone who touches her shall be unclean until evening. Anything on which she lies or sits during her impurity shall be unclean. Anyone who touches her bed shall wash his garments, bathe in water, and be unclean until evening" (15:19).[4] Some fifteen centuries later, the Catholic bishop Theodore ordered that "women shall not in the time of impurity enter into a church or communicate; if they presume to do this they shall fast for three weeks."[5]

In Saudi Arabia the sexes are rigidly segregated because of men's belief that women cannot resist the temptation of sex and will submit automatically to the erotic overtures of any male. Muslims consider that the lifelong sexual fidelity of a man to a woman is against human nature; but a Saudi woman "must be a virgin when she marries, and must remain faithful to her husband. The honour of her family is at stake. The worst sin she can commit is to permit her sex to be enjoyed by anyone but her husband."[6]

These peculiar myths about women are related to the mother's role as the primary maker or "technologist" of the body. As embryos, we are already being refined in her *bottich;* our incipient tendons and bones are taking shape in response to her structure and movements. Through her we enter the world. Nestled in her arms and crawling at her feet, we learn what it is to be and use a body. She is with us as we find out how to grasp and expel. While she is watching us we try out different ways of moving about. We lift our eyes and begin to survey the world. Playing with her, we find that our sounds can entertain us as well as communicate our needs.

Usually alone with our mother in the home for the first few years of our life, we get a sense of the difference between the bodily comfort associated with the world of women and the bodily striving associated with the world of men. In recent decades many men and women have altered this old division, as fathers have taken a more active role in the birth process and in the care of their infants. But in my family, as in many, gender differences were extreme. My father was the only wage-earner in my home. He did hard physical labor most of his life, often working out of town six days a week in heavy construction. He hunted, fished, played sports, gambled, and drank. My mother did none of these things. She managed the house my father built when I was four years old, and took care of me, my father, her father (who lived next door) and my father's parents (who lived around the corner).

I grew up perceiving my father's world as unpleasant and even dangerous. When I was an infant my father ruptured the meniscus in his knee playing basketball, so seriously that he couldn't be drafted during World War II. His friend once was shot in the hand during deer season. Men we knew were always getting hurt on construction jobs. Going hunting and fishing meant I had to get up at five o'clock on freezing mornings and slush through water and mud to spend hours waiting for ducks or trout to show up.

My mother's world seemed safer and more comfortable. As a boy I developed such extreme asthma, and frequent illnesses, that I had to retire to that world. Since I couldn't run for a block without getting violent spasms, I could hardly wade for miles through muddy fields shooting pheasants, or run bases fast enough to be a good ballplayer. So I spent years of nonschool time alone in my room, drugged on Sedatol and Benedryl and reading books, listening to the radio, spinning out endless fantasies. Every summer my mother and I went off by ourselves to the Monterey Bay, where our friends were other mothers, many girls, and hardly any boys. The fathers appeared only for short weekends.

Although the female world seemed safer than the hurly-burly world of my father, it also seemed stifling, like asthma. There was no room for movement and exploration. I felt trapped in it, afraid to venture out into what I was learning was the "man's world."

During puberty I left the woman's world, and stayed away for the next twenty years of my life. I entered the exclusively male Christian Brothers' High School, then the all-male Santa Clara University. My girlfriends were from girls' schools. When I graduated from college in 1955, I went to work for Westinghouse, and visited several of their factories as part of a management training program. The only women I knew there were secretaries. And in 1956 I plunged into the completely segregated world of Jesuits, where the only women I saw for several years were the mothers and sisters of fellow Jesuits, who visited during carefully regulated hours. I taught philosophy for three years at what was then the exclusively male Loyola University of Los Angeles. Even when I had left the Jesuits and was finishing my Ph.D. in philosophy at Yale, I never had a single female teacher (there was only one on the philosophy faculty); Yale College was still exclusively male.

Predictably, my perceptions of women were obscure. Women seemed like foreigners — exotic strangers speaking an unintelligible language, totally unlike myself. I did not see a vagina, not even a photograph of one, until I was thirty-three years old, the first time I had sexual intercourse. I didn't even know what intercourse was before I entered college. You may find it hard to believe that I went through adolescence thinking that babies were born through belly-buttons and were conceived by some vague rubbing together of the man's and woman's pelvises in mutual masturbation.

When I finally reentered the two-sexed world, I slowly came to realize how scarred I had been by those cloistered years. I found it virtually impossible to maintain an intimate relationship with a woman as an ordinary adult human being like myself. I felt I constantly had to be on the watch that I was not losing my power. My old feelings of being stifled and trapped within the world of women erupted so violently that they caused conflicts in my professional life, with such women as Ida Rolf and Thérèse Berthérat, and disrupted my marriage.

I'm embarrassed to admit that even now I feel that women are more dangerous than men. I know, of course, in my waking state that the Krishnas and Weinbergers are destroying the planet. But their danger seems almost simplistic, like the threat posed by a skilled handball player. I have to fight against their destructiveness, but the nature of the fight is clear to me. But women with whom I have been intimate can throw me into depression and inactivity with a glance or a word. I find myself lost in those interactions, trapped as I felt in my childhood home, playing with my toys while my mother puttered around the house. Because I don't know how to get out, I feel impotent.

It's not too hard for me to understand from this frustrated vantage point why I, along with many other men and women, have opted for ideologies that attempt to free "mind" from "body." The solitary role of women, like my mother, in the early teaching of the bodily arts to infants elicits an ambivalent response. On the one hand, we become addicted to the sensual pleasures associated exclusively for most of us with being held and nurtured by a woman's body. On the other hand, that addiction inhibits our movement into the adult world of freedom and clarity. Our unbalanced social arrangements teach both men and women that we can define ourselves as free human beings only by entering into a painful struggle to extricate ourselves from the pulls of the flesh, the maternal comforts of body warmth, food, caresses, soothing sounds, ravishing smiles and glances.[7] The "technology of alienation" has seemed to me, as it must seem to others, judging from its success, an effective way to disconnect myself from what I learned were the "bribes" of flesh.

Dorothy Dinnerstein describes the kind of self I was trained to think I had to create, one that stands clear of the flesh to maintain a perspective on it:

> "I"-ness wholly free of the chaotic carnal atmosphere of infancy, uncontaminated humanness, is reserved for men. And the integrity of this artificially pure, artificially simplified male humanness is preserved by projecting the magical charms and joys of the body, and its mucky, humbling limitations, onto the ambiguous goddess of the nursery."[8]

That self is the soul that the Platonists and the Vedic philosophers say existed for all eternity in some pure, nonfleshly realm: Christian theologians pronounce it created directly by God. The body, however, comes from the mother; she is the one to blame for our physical pain and for what wise men have labeled the "curse" of death. With her deft fingers and soft voice, she creates the illusion that making love, caring for infants, and eating apples are more interesting than war games and monument-building. To allow ourselves to believe that seductive illusion would entail making radical changes in our world.

"Good," argue male philosophers and theologians, comes from man's joining with the father god or the demiurge to shape the primordial chaos through science and technology: ordering the wild forests into pasturelands, dividing the relentless flow of time into controlled intervals of work, rest, and play; mapping space into national grids jealously guarded by armies. "Evil" and "error" in these popular ideologies come both from the body and from women like Eve, Pandora, and Circe, who lure simple men toward self-destruction.

Some male theologians chose one woman, Mary, to remove the taint placed on our souls by Eve, and created an account of her victory that would discredit the sensuality of her sisters. She conceived, they say, without the insertion of surging penis into wet vagina, and gave birth without having the infant pass through her cervix. With her hymen intact, Mary represents a safe heroine for the chevaliers of all times; for it is through the rupture of that hymen, argued Augustine, that all evil passes into humanity — Pandora opening her box.

The ultimate human achievement — heaven or nirvana — is often conceived of as a transcendence of sexuality, a state in which we become like angels, "liberated" from the relentless tide of physical entropy. Such an ideal is appropriately expressed in the traditional image of the diamond: clear, indeed, but impenetrable, lacking organic movement, a Cartesian mind in pure form. Heroes in the quest for a liberated self are like Ulysses, Percival, and John Wayne, wandering the earth seeking ways to transcend human weakness. Their devoted wives tend the hearth, brewing soup for them to drink when they return home tired from their battles and from spending their seed with more exotic sirens. Healed by their wives, they plunge back once more into their "vision quests."

We men have often chosen to live in what we like to think are the spiritual highlands, where the air and streams are pure, the views long. Women seem to us to live in the swamps, where every kind of fungus breeds with ease, nourished by abundant moisture and warmth. Both men and women fear the turbulent murkiness of matter which might, like quicksand, suffocate us. The cool definitions of patriarchal systems and mathematical grids are refreshing respites from fleshly confusion.

Women still do what they have always done; it is men, according to these myths, who have created what are said to be the advances in civilization. They have evolved from hunters and farmers to philosophers, theologians, painters, sculptors, politicians, scientists, and bankers, gaining increasing mastery over unruly natural impulses. Freud argued that women actually retard the advances in civilization:

> Women represent the interests of the family and of sexual life. The work of civilization has become increasingly the business of men, it confronts them with ever more difficult tasks and compels them to carry out instinctual sublimations of which women are little capable. Since a man does not have unlimited quantities of psychical energy at his disposal, he has to accomplish his tasks by making an expedient distribution of his libido. What he employs for cultural aims he to a great extent withdraws from women and sexual life. His constant association with men, and his dependence on his relations with them, even estrange him from his duties as a husband and a father. Thus the woman

finds herself forced into the background by the claims of civilization and she adopts a hostile attitude towards it.[9]

Given the hold these myths have on our guts and neural pathways, it's not surprising that authorities in every kind of society have worked to keep the role of women carefully defined and controlled. Fascists, Communists, Muslims, Hindus, Catholics, and conservatives in our own society have been united on the necessity of keeping the "wife" confined to her solitary role in the home. Hitler wrote that

> it is the man's and woman's duty to preserve man himself. In this most noble mission of the sexes, we also discover the basis of their individual talents, which Providence, in its eternal wisdom, gave to both of them immutably. Thus, it is the highest task to make the founding of a family possible to the mates in life and companions in work. Its final destruction would mean the end of every form of higher humanity. No matter how far woman's sphere of activity can be stretched, the ultimate aim of a truly organic and logical development must always be the creation of a family.[10]

Despite a theoretical affirmation of sexual equality, women rarely appear at the centers of power in Communist countries. Russian feminists have been arrested and jailed as political dissidents. Post–Revolutionary Russian women now have a full workload as well as their traditional jobs as housewives.

Conservative politicians in the United States have taken militant stands against the advances of feminine power. Phyllis Schlafly, a leader of the "pro-family" movement, writes: "I'm a Christian and all good Christians believe that women are special and that God made men to take care of women, to protect them and to go to war for them, to help them with their jackets and make sure nobody else messes with them."[11] The director of the Conservative Caucus argues that the "attack on the family" began when women got the right to vote, because that represented "a conscious policy of government to liberate the wife from the leadership of the husband. It used to be that in recognition of the family as the basic unit of society, we had one family, one vote. And we have seen the trend instead to have one person, one vote."[12] With few exceptions, Christian churches have staunchly opposed the ordination of women as ministers or priests. In a recent address to an international congress of nuns, Pope John Paul said that nuns have the most supreme of women's vocations: to serve the work of priests.

These various social controls and the ideologies that justify them give us a sense of security in a shifting world of emotions. In relation to these rigidly segregated roles we get some idea of how to sculpt our adult identities. But the persistence of this network of sexual imbalances is an increasing threat to our bodily existence. Dinnerstein writes that

the proverbial forms of emotional interdependence between men and women, far from being a guarantee of human survival, are at this point an active menace to it. Together, cozily interdependent male and female quasi-adults are letting the fate of our species slip through their fingers: the males — who make public policy — through inadequate emotional contact with survival-essential considerations; and the females — who have better contact with these considerations — through inadequate authority to make public policy."[13]

WOMEN, EMPIRICISM AND HEALING

These contrasts between popular myths about women and the realities of destructive male power have been the subject of many psychological and sociological studies during recent years. I want to highlight how these contrasts are reflected in two different approaches to shaping our bodies.

The technology of alienation has been a male achievement. Male theologians, philosophers, and scientists have articulated the dualistic theory undermining the value of sensual authority, and male mystics have developed techniques for suppressing the impulses of the body. Until the postwar era, males were the officially sanctioned healers. Males created the industrial and commercial world that shapes most of us from day to day; male politicians and military officers have evolved the strategies of mass slaughter.

What I have called the technology of authenticity has always run quietly along next to the other, sometimes entering into open conflict with it. It evolved from the home and the ways in which women give birth and care for their children.

The two technologies are based on different kinds of evidence and support radically different ideas about authority. The technology of alienation is based on "mediated empiricism"; that is, on data received through instruments and codified into mathematical language. Its prototype is physics, the ideal toward which other kinds of empiricism strive. It was created by male scientists such as Galileo and Newton to explore the macroscopic and the microscopic, which are outside the range of sensory perception. It has to be mediated by instruments, because the observer is incapable of immediate contact with subatomic particles, cells, and stars. It is also mediated in that it admits only those aspects of data that can be expressed in mathematical language, which is thought to reduce human fallibility. The connections between empirical knowledge expressed mathematically and the perceptible world are expressed in often conflicting and highly complex epistemologies.

Mediated empiricism is stereotypically male. It was originally created by men. Its contemporary leaders are predominantly male. Abstract ways of reducing sensory data and fascination with elaborate instrumentation have been characteristically male.

Not surprisingly, this kind of empiricism produces the same model of authority supported by prescientific religious dogmatism — in different garb, but with the same sex, the priesthood of male scientists. Authoritative knowledge belongs exclusively to "experts," those few people who possess mathematical knowledge, understanding of instrumentation, and familiarity with data in a given field. That kind of knowledge is not even accessible to most scientists, let alone an ordinary person.

The technology of authenticity is based on data that are immediately accessible to our senses; it represents an "immediate empiricism." A prototype is natural science: Aristotle studying the species sent him by Alexander the Great, Darwin ferreting out varieties of flora and fauna on the Galapagos, or Jane Goodall studying the secret lives of Tanzanian chimpanzees.

Immediate empiricism suggests a different model of authority. "Experts" are those who pay attention, who open their eyes. The data are readily accessible to the senses, and their patterns are expressed in descriptive language. Although such research requires dedicated effort, as with Freud, who spent years in his salon listening to and observing his patients, or Margaret Mead, who traveled to remote islands in the Pacific, any ordinary person can understand the methods used.

Another prototype for this kind of empiricism is a type of healing developed primarily by women to assist each other in giving birth, raising children, and caring for the sick and dying. Dinnerstein writes that the bodily changes associated with childbearing give women an experience "which means feeling through in a peculiarly primitive and intimate way what it is to be human: to be knowingly part, that is, of a process that started before we were born and continues after we die. Humanness itself, then, is in this particular sense more firmly forced on women than on man."[14]

Women traditionally have been trained in the skills necessary to bring their infants through the many dangers of the early months. Mothers' hands, sensitized to new flesh, easily give massages, soothe the sick, relieve pain, and comfort the dying. Women are expert at diagnosis, having learned to detect the subtle shifts in the smell of their infants' excretions, the temperature of their bodies, or the color of their skin that signal the onset of an illness. Required to attend to the unspoken needs of their infants, women develop the ability to notice things about the body that often escape men's awareness.

The French obstetrician Frederick Leboyer has published a book illustrating the sophisticated massage practiced by ordinary East Indian mothers on their children,[15] and a British anthropologist has described infant-handling techniques practiced by West Indian women living in Great Britain. The sophistication of the wide range of techniques employed by these women rivals that of the most advanced schools of body therapy.[16]

With direct access to the birth process, women are more readily adept at midwifery than men are. Herbalism developed within women's culture as a part of cuisine, as women were trained in the refinements of smell and taste necessary to experiment with various combinations of food.

The strange fact is that such simple methods of healing have evoked extreme measures of repression, even slaughter.

The history of European witchcraft is one example. You may have grown up, like me, with images of witches as old hags bent on hexing little children and innocent young maidens. But recent research has revealed that our popular notions are largely stereotypes that reflect our fears rather than reality. By and large, witches were women healers who practiced the laying on of hands, herbalism, and body-reading. They were a threat to established authorities because, like Galileo, they challenged popular dogmas.

The witch-healer's methods were as great a threat as her results, for the witch was an empiricist: she relied on her senses rather than on faith or doctrine, she believed in trial and error, cause and effect. Her attitude was not religiously passive, but actively inquiring. She trusted her ability to find ways to deal with disease, pregnancy and childbirth . . . It was witches who developed an extensive understanding of bones and muscles, herbs and drugs, while physicians were still deriving their prognoses from astrology and alchemists were trying to turn lead into gold. So great was the witches' knowledge that in 1527, Paracelsus, considered the 'father of modern medicine,' burned his text on pharmaceuticals, confessing that he "had learned from the Sorceress all he knew."[17]

Beginning in the fifteenth century was a three-hundred-year period of persecution throughout Europe, during which one writer estimates that some nine million women were put to death; another puts the figure between 30,000 and several million.[18] In some villages in France one woman was preserved, not because she was judged innocent but simply because she was the last woman. Such persecution bears striking resemblances to the Holocaust. Both were centered in Germany; the numbers of people killed were staggering; and the motivation for the slaughters were similar: The Inquisitors and the Nazis played on popular fears of sexual powers supposedly held by women healers and Jews.

The witch trials brought to light the ideological unity among Church, State, and the new biomedical experts. In 1486 two German Dominican priests wrote a catechism of demonology, the *Malleus Maleficarum* ("The Hammer of Witches"). It says: "And if it is asked how it is possible to distinguish whether an illness is caused by witchcraft or by some natural physical defect, we answer that the first way is by means of the judgment of doctors."[19] During this period the Church legitimized male physicians' exclusive claims to the right to heal, denouncing nonprofessional healing as equivalent to heresy. According to the *Malleus*, "If a woman dare to cure without having studied she is a witch and must die." But there was no way open for women to study.[20]

Where scientific rationalism began to replace Roman Catholic theology as the dominant ideology, people were taught to think of satanic witchcraft as primitive superstition. The kinds of healing practiced by women were considered remnants of prescientific consciousness, no longer worth taking seriously. Male scientists linked their trances and visions with the newly discovered "disease" hysteria, associated with having a uterus.

The power of women over the well-being of the body, formerly undermined by priests and magistrates, was now to be attacked by a cadre of gynecologists, obstetricians, psychiatrists, and pediatricians. Women's usurpation of authority was the underlying theme that united the different rationales of control employed by the medieval Church and the new medical establishment. In the first case the popular Christian view associated female rebelliousness with the devil. In a supposedly more enlightened secular age, women's rebellion was said to be caused by deficiencies in female organs. Ovariotomies, clitoridectomies, and hysterectomies replaced the racks and water stools of the Inquisition. An 1880 American Medical Association *Journal* article describes surgeons rushing to practice abdominal and pelvic surgery on women, "as restless and ambitious a throng as ever fought for fame upon the battlefield."[21]

While gynecological theory was being born, Freud was studying in Paris under Charcot, whose Salpetrière Clinic was reminiscent of the ecclesiastical courts of earlier centuries. The notes on a particular session describe Charcot's work with a woman displaying hysterical spasms. He suspends the attack "by placing first his hand, then the end of a baton, on the woman's ovaries. He withdraws the baton, and there is a fresh attack, which he accelerates by administering inhalations of amyl nitrate. The afflicted woman then cries out for the sex-baton in words that are devoid of any metaphor: B. is taken away and her delirium continues."[22]

The ideology of hysteria was also used to discredit women's authority in the area of birth. Because of their supposedly emotional nature, women were forbidden entry into medical schools. But obstetricians organized to pass laws that required people overseeing births to know how to use forceps and knife — knowledge available only in the medical schools. Midwives were seen as an obstacle to the spread of the new science. In the United States between 1900 and 1930, they were almost totally eliminated.[23]

During this same period, however, a shift was occurring that would encourage people to take a fresh look at supposedly irrational modes of knowing. Male thinkers themselves began to call into question the validity of mediated empiricism.

I have called your attention to one indication of that shift in Chapter Three, where I outlined the new understanding of how our perceptions are affected by our subjective images of our bodies. On other fronts, theoreticians such as Kurt Gödel and Bertrand Russell showed that the supposedly universal mathematical language of science was built on presuppositions that could not be proven but were only assumed to be true. Phenomenologists such as Edmund Husserl and Maurice Merleau-Ponty showed how empiricism's lack of connection with its origins in the perceiving body had led to extreme relativism, and physicists such as Niels Bohr and Albert Einstein exposed the tenuous foundations on which classical theories of matter had been constructed.

A theoretical framework that made it possible to reconcile the ancient conflicts between male and female stereotypes and the kinds of knowledge associated with each began to emerge. For the first time in modern history there arose two movements, intimately related, whose leaders were predominantly women: modern dance and somatic therapy.

Isadora Duncan and Mary Wigman broke with classical ballet's molding of dancers into idealistic forms designed by male choreographers. These and other women, along with men such as Rudolf Laban and Merce Cunningham, created an aesthetic based on the elaboration of movements coming from impulses within the bodies of the dancers themselves. In contrast to classical ballet, which rewarded dancers according to their ability to shape themselves in congruence with outside forms, modern dance encouraged self-awareness and flexibility of expression.

The populist, nonauthoritarian dimension of this movement is found in the works of such people as Deborah Hay and Mary Whitehouse, who use nondancers in their events, asserting the beauty of ordinary movement done with simple awareness. In a dance designed by Hay, for example, thirty ordinary people walked out on the stage three consecutive times, carrying chairs, placing the chairs, sitting down, getting up, and walking off. The work made the audience aware of the complex beauty of the simple movements that we perform every moment of our lives.*

Another instance of the new aesthetic is the work of Pilobolus, a dance collective appropriately named for a fungus that grows on horse manure. The dancers seem to delight in the most unconventional sorts of movement, but they dance with each other, forming strange configurations that look like creatures from another planet or animals in fairy tales. These are not alienated individuals striving to create predetermined external forms, with a hierarchy of stars and chorus. They seem instead like interacting parts of a consensual organism.

Closely related to modern dance is what Thomas Hanna has named somatic therapy, a variety of methods aimed at the health of the body. Women also predominate here: Elsa Gindler, Bess Mensendieck, Ilsa Middendorf, Lillemore Johnsen, Gerda Alexander, Ida Rolf, Gerda Boyesen, Charlotte Selver, Jean Ayres, Emilie Conrad, Judith Aston, and so on. The technology they employ bears curious resemblances to the old witchcraft. In contrast to the complicated diagnostic and therapeutic methods of scientific medicine, somatic therapies are based on immediate empiricism. Like modern dance, their practitioners are people who are skilled at perceiving. Diagnosis is done by looking, feeling, and listening. Healing is simple: sometimes it is accomplished by manipulating muscles and limbs; at other times, by guiding people's awareness to specific areas of their bodies or teaching them basic forms of movement and exercise.

---

*What I'm saying applies only to what might be called the avantgarde of modern dancers. Martha Graham, Murray Lewis, and others have crystallized the aesthetic into a new ideal form.

The discoveries that led to the creation of this technology were based on a return to the obvious, an acknowledgement of the authority of perception unmediated by instruments and mathematics.

Elsa Gindler, for example, had to deal with tuberculosis which she contracted in Berlin in 1906. Her physician told her to go to a sanatorium and predicted she would die. Refusing to accept his diagnosis, she spent six months directing her awareness to the movement of breath in her left side. Her lungs healed, and she became a pioneer in teaching others the curative powers of simple awareness, a method now commonplace throughout Europe and the United States. She writes of her work:

> We see to it that our students do not learn an exercise: rather, the *Gymnastik* exercises are a means by which we attempt to increase intelligence. When we breathe, we do not learn fixed exercises, rather exercises are the means of our getting aquainted with the workings of our lungs, either through inducing or releasing holdings. When we become aware that our shoulder-girdle is not in a position where it works easily, we do not put it into the correct position from without. That does not really help anything, for as soon as the person is busy with something else he forgets his shoulder-girdle... We carefully examine it as to detail of form and usage. With the help of a skeleton we find out how it can best fulfill its function. We compare our functioning with that of the skeleton and then work to find out what has to happen within ourselves to come closer to such functioning.[24]

On a camping trip in the Rockies in 1917, after her graduation from Barnard College, Ida Rolf nearly died of pneumonia. She was saved by a simple manipulation of her spine by a small-town osteopath. She devoted the subsequent sixty years of her life to exploring the results of touching people in simple and direct ways. Since she had fought for some fifty years to gain even a modest hearing in the male world of science (she got her Ph.D. in biochemistry at Columbia in 1920), I was not surprised that in her late seventies, when I worked with her, Dr, Rolf often behaved like a cantankerous authoritarian. Despite her bossiness and fixed ideas, she constantly trained me to be my own authority. She insisted that I pay attention to what I saw and felt in another's body, and to base what I was doing on the results of what I actually observed, not on what she or others said.

As might be expected, pioneers such as Rolf and Gindler frequently ran into conflict with established authorities. Elsa Gindler refused to give her work a name and to form an identifiable school, because that would have meant that she had to include a program of Nazi indoctrination in her training. During the war she gave classes for Jews living secretly in Berlin; a week before the Russians liberated the city, a Nazi youth hurled a fire bomb into her studio, which killed her Jewish students and destroyed all her writings. Ida Rolf always lived on the margin of the law. Many of her professional associates were put in jail at one time or another, and records of their research were confiscated. One of her personal healers was jailed in Los Angeles as recently as 1978. Her institute operates outside the law, subject to persecution for practicing medicine without a license.

The methods developed by these women, and by men such as F. Matthias Alexander who shared their approach, suggest a way of recovering our sensual authority. Their techniques teach people how to heal themselves by reconnecting with their bodies. These people had the courage to assert the forgotten obvious: that we have more immediate access to our own bodies and perceptions than any scientist or instrument; all we need do is revive our skills of paying attention. You've already had tastes of that technology in responding to my various invitations to reflect on your bodily behavior. The next two chapters are dedicated to an elaboration of that technology and an exploration of its social implications.

CHAPTER EIGHT

COMING TO OUR SENSES

The first six chapters of this book are dedicated to diagnosing the "mind-body" fracture and its implications for our personal lives and social policies. The previous chapter begins to suggest a cure. The very stereotype "woman" contains humanity's store of knowledge about how to reconnect with our sensual experience. Women forced into that stereotype have pioneered the way back, and this chapter is devoted to the details of that healing path.

After twenty years of unsuccessful attempts to cure her back pain with surgery and drugs, Sister Ruth encountered the technology of authenticity. Her back pains disappeared. Her methods might seem ridiculously simple compared to the sophisticated and expensive methods she had been subjected to in previous years: She began walking two miles each morning, she learned to play racquetball from a teacher who emphasized self-awareness rather than the "right way" to play, and she had a brief series of sessions with me, during which I used my hands to show her how to lengthen certain muscles, allow more movements in specific joints, and breathe more fully.

What Sister Ruth did was not particularly significant; she had done similar things in the past without much success. The significance lay in the manner in which she did them. She used these three activities to reconnect with her own resources for healing herself, for taking charge of her own power. In the past, she had obediently adhered to the prescriptions of male experts. She did not interpret their failures as spurs to explore other sources of knowledge. With fatalistic passivity, she assumed that her only alternative was to await medicine's discovery of more sophisticated prescriptions to fix her stubborn body.

Had you asked me twenty years ago if I accepted the mind-body split, I would have vigorously denied it, barraging you with a wide range of intellectual arguments from Darwin, Freud, Wittgenstein, Heidegger, William James, and the New Testament. But at that time I was a Jesuit celibate who lived with a vow of obedience to the Pope and a state of chronic back pain. Any person observing my life without listening to my words would have seen me as an embodiment of an ideology that subjects sensual experience to reason and official authority.

I doubt that many of you would admit to a belief in mind-body dualism. It is an idea whose time has passed, which is roundly discredited in every field of thought — even in theology, once its most generous patron. But now that this book has accustomed you to seeing belief systems strutting down the street, you will realize that dualism is the way in which we are trained to move, perceive, and feel. The belief is generated from and supported by this training. We can't loosen its grip on our muscles simply by rational analysis; a true transformation requires practical strategies that help us to recover a sense of our own authority. I have called such strategies the technology of authenticity.

> *Authenticity* comes from the Greek and originally meant to do something oneself, to have a sense that one's actions and feelings are one's own. A person who truly expresses himself or herself through feelings and thoughts is considered "genuine" or "authentic," in contrast to the hypocrite, who is alienated from what he or she expresses.

The technology of alienation accustoms us to sense a void

between "I" and my flesh, and between "I" and "you." Because we are led to feel that we are not in immediate contact with the palpable world, we sense that we need experts who understand that world enough to tell us what to do. Feeling that we are disconnected from each other, we require outside mediators to resolve our inevitable disputes.

Authenticity, on the other hand, requires us to take authority over our lives. The authentic person is the one who is aware that his or her worldly behavior is indeed his or hers. My actions have consequences for which I am responsible. Hurting others, failing to pay attention to certain groups of people as fellow humans, recoiling from acting when I perceive I should: these are *my* actions.

The fundamental shift from alienation to authenticity is deceptively simple: it requires diverting our awareness from the opinions of those outside us toward our own perceptions and feelings. I say "deceptively" simple because those of us who have received a lifelong education in forgetting our senses need a long and arduous process to recover the simple abilities of touching and seeing. One teacher of the new technology points out:

> When people ask me why my work looks so simple, I say it took me many, many years. When I tell them that I've been searching and experimenting for fifteen years and I think I'm just beginning to understand something, they're astonished: "Fifteen years?" And I say, that's not very much. And it's as if they would like to see something very dramatic immediately, and they are not trained to see the process of touching and waiting and the beauty of this simplicity.[1]

It is easy to take such skills for granted, just as we often take for granted the similar skills possessed by a woman who has been successful in raising her children and managing her home.

Pause

You may be wondering, "What can I do about all this? You've raised provocative questions, but how can I do something to resolve them?"

Take the example of sitting. I assume that by this point in the book you've gathered a good amount of data about how this technique operates in your life. You know how you've learned to sit. You have some sense of characteristic tensions associated with certain postures. You have learned how to make yourself more mobile in your chair and how to relax your back and support your head. You've probably had some insights into how to vary the kinds of seating to which you have accustomed yourself. You have already bridged the gap between yourself and sitting, the alienation between you and seats. You are on the way toward becoming your own authority in this realm, capable of evaluating what the experts say about it.

Contemplate a more complex issue — a chronic pain you or a loved one might have: arthritis, bursitis, neck or back pain, tension headaches. Consider what might be required for you to become your own authority in dealing with such a problem. If you are not to become a pawn of traditional or avant-garde medical experts, you have to pay a great deal of attention to shifts in that pain related to exercise, emotional stress, habitual posture, sexual experiences, and diet. You also need to know some basic facts about anatomy. Against that backdrop you will probably have to try out different expert prescriptions, observing the results of such antidotes as pain-killers, a new kind of exercise series, manipulative strategies, or relaxation techniques. As you do all these things over a period of months or years, you will undoubtedly encounter psychological barriers to pursuing the process, perhaps related to old feelings of personal inadequacy. You might run up against social barriers; for example, you might discover that your back pain is directly related to the repressive atmosphere in your office, and that to cure it would mean raising questions about your source of income.

The technology of authenticity is serious business, involving far more than what one might label a return to common sense (a term that has been used to justify the most extreme instances of fanaticism and bigotry). The creators of this technology are a group of experimentalists whose rigor and courage in pursuing the implications of their research match that of our pioneers in natural science. Nikolas Tinbergen, in his Nobel Prize address, compared the field to his own field of ethology.[2] Although the original creators of the technology were spread over an area ranging from Australia to Eastern Europe and were usually ignorant of each other's work, they reached a remarkable unity in creating an approach designed to teach people how to reclaim the authority of their sensual experience. The methods are inherently democratic, simple, and direct, and expressed in a language accessible to any ordinary person.

Because the technology arose outside the world of universities, laboratories, and books, those familiar with it tend to concentrate on its practical applications for personal health and well-being — expanding the "human potential." Both its supporters and its critics tend to overlook its social and philosophical significance. Mostly upper- and middle-class whites pay large sums of money to study the techniques in resorts like California's Esalen Institute or Maine's Monhegan Island, far removed from urban poverty and international terrorism. They often become devotees of a specialized school: Rolfing, bioenergetics, the Alexander technique, Lomi work, neo-Reichian therapy, and so on. Such schools typically become sects with rigid hierarchies of authority, ideal bodies, and ambitious commercial ventures.

But I was originally drawn to studying this new approach to the body because of its implications for social change. It was the 1960s and I was actively engaged in resistance to the Vietnam War. Herbert Marcuse and Norman O. Brown were arguing that revolutions would continually recycle people — old revolutionaries becoming the new dictators — unless there were radical changes in human beings, specifically in their bodies. While writing my doctoral dissertation at Yale on the relationship between the body and the body politic,

I asked myself why I was not engaged in reforming flesh instead of thinking about it, and plunged into a ten-year study of the new technology. This book and my earlier book, *The Protean Body,* arose from my desire to show the little-recognized social implications of these apparently banal methods.

Moreover, as I began to learn more about the lives of the various somatic pioneers, I was struck by how our emphasis on the variety of techniques they developed obscures their unity in a radical commitment to the authority of sensual experience. Reflecting similar tendencies in history to institutionalize the creative insights of people such as the Buddha, Jesus, Marx, and Freud, popular emphasis is usually placed on differences among various somatic methods. Practitioners are trained to imitate the "moves" developed by the founder without being prompted to explore the radical conversion he or she underwent. Clients are often encouraged to learn the techniques but not educated to recover their own sensual authority. I came to realize that an essential ingredient in understanding the new technology, as distinct from its variety of techniques, is to study the lives of the founders. Before they became publicly recognized experts with defined methods, they all experienced similar conversions from alienation to authenticity. During those events they learned valuable lessons about how we all might reconnect with our bodily wisdom.

A convenient date to mark the earliest formulations of the technology of authenticity is 1890, when F. Matthias Alexander cured himself of laryngitis. In that year Elsa Gindler was beginning to teach physical education in Berlin, shortly before she contracted tuberculosis; Isadora Duncan, Mary Wigman, and Rudolf Laban were beginning to perform new kinds of dance; Bess Mensendieck was about to leave New York for Hamburg, where she developed the method of posture training that later became a required course in Ivy League prep schools and universities; and Wilhelm Reich had just been born in Galicia and Ida Rolf in Manhasset, Long Island.

Here are the stories of four somatic pioneers, I am focusing not on the particular methods they developed but on their shared recovery of sensual authority.

(1)

F. Matthias Alexander had undertaken a career as a Shakespearean actor in Melbourne, Australia, when he began to have trouble with his voice. At first he was told that his breathing was so heavy that people could hear sucking noises while he was reciting. Then he began to suffer from laryngitis, which became so serious that he couldn't use his voice in performances. Voice teachers were unable to help him. He visited several physicians, two of whom suggested surgery to shorten his uvula, the small piece of soft tissue toward the rear of the roof of the mouth, which they judged to be too long. He refused.

Alexander was puzzled when he noticed that although he could talk for hours with friends without experiencing the least discomfort, a few moments on the stage left him hoarse. Wondering what the difference was between his ordinary conversation and his performances, he spent several months working in a studio filled with mirrors, trying to re-create what he did when he was about to go on stage. He gradually discovered a complex system of attitudes, feelings, and muscular patterns which he called his "habit of performance." That habit contrasted with the way he spoke with friends. When he was preparing to play a role, his self-consciousness manifested itself in a peculiar tilting back of his head, accompanied by tension in his neck muscles which strained his larynx and inhibited his breathing. None of these tensions were present when he was conversing with his friends.

Alexander expressed his fundamental discovery in words that many subsequent pioneers echoed:

> I must admit that when I began my investigation, I, in common with most people, conceived of "body" and "mind" as separate parts of the same organism, and consequently believed that human ills, difficulties and shortcomings could be classified as either "mental" or "physical" and dealt with on specifically "mental" or specifically "physical" lines. My practical experiences, however, led me to abandon this point of view and

readers of my books will be aware that the technique described in them is based on the opposite conception, namely, that it is *impossible* to separate "mental" and "physical" processes in any form of human activity.

This change in my conception of the human organism has not come about as the outcome of mere theorizing on my part. It has been forced upon me by the experiences which I have gained through my investigations in a new field of practical experimentation upon the living human being.[3]

The essence of the Alexander Technique came from Alexander's struggles to understand the difference between conversing with friends and reciting on stage. He discovered the difference between what he called the "means whereby" and "end-gaining." When he was with friends, he was simply *there,* responding to the situation without any particular goal. As soon as he got ready to recite his goal took control, evoking the set of muscular rigidities he called his habit of performance. He could not overcome that pattern until he inhibited his obsession with the goal of his activity, a device that inspired my use of "pauses" in this book.

Alexander applied the notion of inhibition to a wide range of human activities, from sports to nuclear war. In the case of the golfer who keeps missing the ball, for example, he argued that it is of little use to keep telling the golfer to keep his eye on the ball. In fact, such piecemeal directions may make his game worse, because they will encourage the golfer's preoccupation with the goal and thus distract him from an awareness of how his body is performing the act. The golfer's teacher has to get a sense of the mechanical pattern that appears in all the golfer's activities and then teach the golfer how to inhibit those ends, through verbal directions, touch, or simple exercises designed to help him experience the unity of his organism. The strategies of inhibition, of course, are the essential and most difficult part of Alexander's work, because they must be able to trick us out of our oldest and favorite ways of doing things.[4]

Alexander applied these ideas to a proposal for a freeze on nuclear weapons as early as 1946. He argued that nations are like the poor golfer or the hoarse actor, dominated by reactions that do not serve their present goals. Stunned and bewildered by the Frankenstein monster they have created, nations do not have a sense that they can prevent irreparable harm to the world. To succeed in inhibiting their mechanical rush toward destruction,

> we shall be forced to come to a FULL STOP. This may well prove to be the most difficult and valuable task man has ever undertaken until now, for he has gradually been losing a reliable standard of control of reaction, and the ability to take the long view, in his efforts to improve his conditions when he is faced with the need for changing habits of thought and action. This should not surprise anyone who remembers that in most fields of activity man's craze is for speed and for the short view, because he has become possessed by the non-stop attitude and outlook: he is a confirmed end-gainer, without respect to the nature of the means whereby he attempts to gain his desired end, even when he wishes to employ new means whereby he could change his habits of thought and action.[5]

Like many of the somatic pioneers, Alexander did not explicitly relate his work to prevailing notions of authority. He did in fact create an ideal notion of the body, and often acted like a pretentious expert, but at the same time he developed a somatic logic that revealed weaknesses in tendencies to locate authority outside ourselves. He began his research by challenging his physician's authority. In his later teaching he found that the desire to please outside authorities was a fundamental source of misuse of the body. In his analysis of the golfer, for example, he discovered that "the greater his desire to obey his teacher, the greater will be his incentive to increase the intensity of his efforts, and it is practically certain that in his attempts to translate this desire into action, he will automatically increase the already undue muscle tension which he habitually employs for the act, thus lessening still further his chances of making a successful stroke."[6] In an essay about physical exercise he criticizes the competitive social model by which men organize their world:

> Surely a university boat race should be a friendly contest between men animated by the sporting instinct. Every one of them should wish the victory to go to the best crew. It should be an experience of pleasure, happiness and healthy recreation to all concerned, not an unnatural struggle involving distortion and loss of consciousness through the "determination" to gain an end even at the cost of personal exhaustion and damage. What difference does it really make in the long run whether or not an Oxford or Cambridge crew wins the boat race in a particular year?"[7]

(2)

In the previous chapter I spoke of Elsa Gindler. One of her pupils, Charlotte Selver, left Berlin for New York in 1939, fleeing Hitler, and her quiet sensory work began to spread throughout the United States with the assistance of such people as Erich Fromm, Fritz Perls, Alan Watts, and Suzuki Roshi. Her work is as simple and direct as the powers we feel we have lost; she merely teaches students to pay attention. During a session a group might spend an hour just getting up and sitting down, noticing the sensory data present in that simple act: shifts in weight, muscular tension, breathing rate, emotional reactions, the quality of surfaces and atmosphere. The work is a methodical training in how to recover the quality of immediacy we had as infants and how to use that quality in the refined way necessary for adult life.

In a passage in which she is explaining why she calls her work "sensory awareness," Selver writes:

> Very often "perception" means only "what I see"
> — it's often very much here in the head. But when one

says sensory, that includes all the senses. The whole nervous system is impregnated by anything which happens, and one must be quiet enough and receptive enough so that this can happen; so that no thought and no word interferes with it. One is just being open for the experience itself. Then, later on, a deep and fruitful thought can emerge, a much more full, verbal expression.[8]

Like many of us, Selver came to these insights in the course of coping with her own body. Gindler's first impression of Selver when she arrived in Berlin in the 1920s was that she was "a beautiful mannequin" who showed no spontaneity in her movements even though for some years she had practiced gymnastics and music. After Selver had studied with her for more than a year, Gindler walked over to her one day, put her hand on her shoulder, and said, "At last, the first movement!" Selver reports: "After that, I began to feel that my movements were hollow, that they did not mean anything. That I was still in the old techniques. It took me a very long time to lose this shellac."[9]

Selver's husband, Charles Brooks, writes of the impact of her work on him:

During my life, I have often rejected one authority only to accept another. Underneath, I was afraid at the thought of living in a world where there was not Someone, somewhat like myself, who *knew*. But I have now come to feel that to know what one is doing with life, it is no use to consult authorities. It is precisely through the veils which authorities have spun for us that our own ears and eyes and nerves must begin to penetrate if our hands are to grasp the world and our hearts to feel it. We must recover our own capacity to taste for ourselves. Then we shall be able to judge also.[10]

(3)

Moshe Feldenkrais was the first person in the West to receive a black belt in judo. While he was a physics student at the Sorbonne in the 1930s, he severely damaged his knee playing soccer. Physicians told him it would never function properly again. Refusing to accept their diagnosis, he got the idea that if he could mimic the exploratory consciousness of the newborn, he might be able to heal himself. During a three-month period he brought himself into a state of inactivity to the point at which he virtually forgot how to use his body. Then, slowly and with meticulous discrimination, he explored the possibilities of movement in each of his joints. He healed his knee and simultaneously discovered a successful method for teaching others to recover their sensual learning capacities. The Felkenkrais teacher is essentially a learner.

In *The Case of Nora*, Feldenkrais describes in some detail how he works with people. Nora was a Swiss woman in her sixties, well educated and intelligent. One morning she awoke and had difficulty getting out of bed; her speech was slurred and she could neither read nor write. After a year in a neurological clinich in Zurich, she had not improved. She was taken home, where it was difficult for her to move about because she could not locate doors and furniture. However, when she was sitting in a chair she conversed normally and intelligently. Nora sought out Feldenkrais, who worked with her for different intervals for over a year, at the end of which she had begun to regain some of the basic functions she had lost: the ability to write her name, distinguish right from left, put her glasses on, locate doors, and walk without bumping into things.

Feldenkrais succeeded by simply paying careful attention to what she did. He describes his first session with Nora: "Examining Nora's head, and gradually reducing the intensity of my touch and of movements for finer appreciation, had an effect on the muscles of her neck, and her head became easier and smoother to move. I felt that she responded very well; her face became alive, her eyes brightened, and her depression disappeared gradually." As he continued to examine the quality of movement in her legs, arms, and trunk, she relaxed further. At the end of this exploration he guided her hand to write the figures 3 and 4 and asked her what she had written. She said "thirty-four," a recognition she hadn't had for over a year.[11]

As the sessions progressed, Feldenkrais's explorations became more subtle, directed in this particular session at trying to learn what precisely was happening in Nora's eyes:

> I tried to identify myself with my patient. What was she actually doing when she intended to read? Where did she intend or expect to read? Did she intend to read the first word of the page, and then fail to direct her gaze, allowing herself to look far into the distance, as though looking at infinity, her eyes not converging at all? Did she see the words she said with her better eye or with both? How on earth was I going to find out?[12]

Notice the radically different model of authority and expertise that is operative in this case. Feldenkrais was expert in his extraordinary ability to allow Nora to teach him what she needed. He was not the traditional therapist, operating with knowledge external to the patient and telling her what to do, but a fellow researcher.

The Feldenkrais method for group education, called Awareness Through Movement, involves explorations of all the imaginable possibilities of moving our bodies in order to get a sense of how to break out of our "habits of performance." It involves returning to childlike methods of finding out how many possible ways there are of walking from here to there.

(4)

Emilie Conrad was a professional dancer in New York for some twenty years who subsequently lived for five years in Haiti studying African dance and native forms of healing. When she returned to the United States, she had a psychotic type of experience, which she describes as a sense of slipping into a crack between the two cultures. She was shocked to find herself neither a New Yorker nor an African dancer.

From that point she saw the effects of culture on our ideas of "body," on our definition of the human form. She writes of this experience: "I had to give up everything I believed. I saw that what I called 'my body' — how I moved, talked, even how I thought — was a cultural imprint. With all my training, I had been teaching 'my body' to dance. But deep inside there was already a dance going on, if I would perceive it — a dance of myriad movement forms beyond anything I could think of. I had to feel it. I had to let it guide me."

After her experience, Conrad worked primarily as a healer, treating people with serious afflictions. In one such case she worked for over a year with a young paraplegic. At the beginning of the work the woman couldn't move her lumbar spine and hips. In the early sessions, which were recorded on film, you can see delicate pulses radiating throughout various parts of the woman's body (somewhat like the orgastic responses Wilhelm Reich investigated) as Conrad places her hands lightly on certain places or gives the woman certain suggestions. By the final session the woman is undulating her spine and initiating sensuous rotations in her hips. She is dancing.

From cases like this and from her own experience, Conrad began to realize that there are subtle levels of movement in our bodies that we ordinarily neglect. When we pay attention to those levels, wonderful things happen: healing often occurs, people find new ways of moving, and frequently their ideas change. "When you pay attention in that way," she writes,

> conventional assumptions about yourself seem to fall away. Since the seventeenth century the body has been thought about as if it were a machine, complete with *pumping* heart and moving *joints*. After the invention of locomotion, the notion of *motivation* became popular. We had to motivate ourselves as if we were static machines that needed an outside source of energy to get us going. We learned to *discipline* the body, *organize* its capabilities, and *harness* its forces into useful chan-

nels. The conclusion that this is what the body actually *is* has placed severe limitations on our sensual life.[13]

Like Selver and Feldenkrais, Conrad has developed exercises, which she calls "Continuum," which do not require any "right" way or ideal toward which a person must move. They are simply explorations. People are taught how to initiate the subtlest possible movements in certain areas of their body, through tiny muscle contractions ("micro movements"), breath, or sound. For several hours at a time people are encouraged to explore the results in their body movement.

My work with Emilie Conrad has had a major impact on the shape of this book. She regularly works at the point at which classical body ideals and organic impulses meet. For example, I might adopt a classical yogic asana but with the instruction to allow any impulses that might arise to dictate subsequent movements of my body. The asana might slowly dissolve into an undulating pattern. It then becomes a key for tapping into otherwise unknown depths of my body rather than an outer form into which I attempt to force my body. It is the same with aerobics: Conrad gets people to use the forms — jogging, forward bends, situps, pushups, and so forth — to elicit more impulsive, idiosyncratic kinds of movement.

Working in this way has helped me to get a much clearer sense of the distinction between using external forms as clues for ways to explore my own capacities for movement and using them as norms for judging myself and others. This distinction is critical in understanding my contrast between the two technologies. Any techniques for dealing with our bodies — ranging from cradling infants to weight-lifting to orthopedic surgery — have neutral value. We can use them to make us more rigid and to train us to denigrate the value of our own experience, or we can use them to refine our ability to discriminate and reconnect with our own power.

AUTHENTICITY AND AUTHORITY

The experience of our bodies, encouraged by the technology of authenticity, suggests that we take a fresh look at some of the basic concepts of social organization.

The dominant values of cultures throughout history have been reflected in hierarchical images of the body, in which one part enjoys preeminence in relation to others. The order differs according to a particular culture's map of the body: The father as head of the family, whose heart is the mother and whose arms and legs are the children; the king or president as the head, the army as arms and hands, the feudal lords or Congress as legs and feet, the citizens as the body; Jesus Christ as the head, Mary as the neck, the Church reflected in Pope, bishops, priests, and people, as the Mystical Body.

The technology of authenticity suggests that the mind is in every cell; no part has superiority, not brain, nor penis, nor vagina, nor heart. Power is dispersed.

The work of the somatic pioneers revealed the significance of every part of the body for every other part. Tension in a person's eyes and neck may be related to rotations in his or her feet. Ulcers, backaches, and anxiety have been traced to systems of stress permeating the entire body. Sensory impulses are not isolated events initiated by discrete sense data that trigger mechanical responses. We grasp sensual patterns *(Gestalten)* which initiate movements (emotions) throughout our entire body, which shape further perceptions in a systematic cycle that enmeshes our body in a world of color, sound, and movement.

Function rather than status determines whether any part of the body takes precedence over others. If you want to go on a hike, your feet and legs become important. If you want to go rock-climbing, you will need strong arms. To read books or do fine embroidery requires sharp eyes. Without a well-functioning gut you will have trouble digesting the food you need for stamina. For a scholar, a broken foot may have no more significance than a few weeks' discomfort; for a professional dancer, it might mean the end of a career.

But apart from such purposes no part can be designated the "metaphysical" ruler of the others; they all retain significance. Power is dispersed in such a way as to profit from each person's unique viewpoint, genius, and range of experiences. This way of experiencing our bodies provides a basis for sensing the interlocking values, for example, of managers and blue- and white-collar workers, or of workers, scientists, and artists.

This is not a farfetched utopian fantasy. Many corporations — hardly the avant-garde in personal liberation movements — are finding that traditional authoritarian models don't produce profits. A consultant for such companies as Atlantic Richfield, General Electric, and IBM writes:

> Now even the large organizations — the last champions of hierarchical structure — are questioning whether the hierarchical structure can fulfill their organizational goals. Many are discovering that the hierarchical method that was so effective in the past is no longer workable, in large part because it lacks horizontal linkages. In the future, institutions will be organized according to a management system based on the networking model. Systems will be designed to provide both lateral and horizontal, even multidirectional and overlapping, linkages . . . [Management] styles will be rooted in informality and equality; its communications style will be lateral, diagonal, and bottom up; and its structure will be cross-disciplinary.[14]

Reflect for a moment on the unique, often disregarded, perspectives that ordinary people have about the institutions in which they live and work. Office workers can see ways of increasing their own productivity that often escape the notice of managers; waiters, waitresses, and bartenders have unique perspectives on how to improve service and increase business; children in a family can often see more clearly than adults what needs to happen to make the family more harmonious; students know how to make schools more effective centers of learning; and nurses and nurses' aides have intimate senses of what patients need. But most institutions are set up so that such voices are rarely heard.

Do you possess a unique body of sensual experiences that would be useful for others but is not acknowledged in your social world?

The technology of authenticity suggests a consensual model of authority in which all members of a community, regardless of where they fit on the old hierarchical scale, have an effective voice in those decisions that affect them. It also suggests a return to a more traditional understanding of expert.

Expert originally meant "an experienced person," one who had felt and perceived a lot, who had tried things out. The notion implied somatic authority.

All of us have some kind of "expertise," derived simply from living in a particular way. Think, for example, of the sensual data about growth, learning, and health a mother possesses from her contacts with her children. Add to this her experiences of working out her emotional relationships with her children, her husband, and in many cases, her employers. Each of us has a treasury of such experiences, which are usually discounted when we think of expertise.

In relation to the questions facing a particular community, some have more experience than others. As a class, the aged have a special claim to expertise regardless of their education or social status, since they have more years of the tasting and feeling that are the ground of intelligence. One might argue that many older people have closed themselves to sensual experience. But I hold that there is a certain inevitability in living. Even a person who has remained apparently closed to seeing and feeling has an enormous amount of data stored in memory, data that can be recalled by a person who asks in the right way. My parents, for example, have been married for fifty-one years. They seem as happy together now as anybody I know, even though their life has sometimes been filled with intense conflicts. Although they don't publish books about how to make relationships work or offer marriage counseling, I suspect that they possess a special kind of knowledge that well-educated psychologists without their success at maneuvering through years of difficult experiences do not have.

Another class of experts are those who have undergone particularly intense experiences. These might include such people as the survivors of concentration camps, Hiroshima and Nagasaki, or near-death experiences like Imelda's. Blacks and Hispanics living in ghettoes, battered women, and abused children have intense somatic experiences that give them a unique and penetrating grasp of the way in which society works.

Those who have deliberately investigated a broad range of somatic experiences constitute another group of experts. These include vagabonds, artists, and explorers. Margaret Mead, for example, was such an expert, not because of her academic credentials but because she traveled to remote corners of the world, carefully watched different kinds of body movement, learned the sounds and rhythms of many languages, and tasted many kinds of food.

Still another group of experts are those who have immersed themselves into a particular world of sensory experience. Painters, skilled craftspeople, and musicians are among those who are adept in one area of experience. Ida Rolf spent her life watching people move, touching thousands of them, and observing the results of her touch. Jane Goodall is spending her life among chimpanzees, observing their habits and learning to communicate with them.

Think of the richness of experience possessed by these various experts, and compare it with that of the people who are designated experts in our society. Our national leaders, for example, are rarely people who have any special claims to authoritative experience. They are generally upper-class Anglo males whose perceptions of humanity have been highly restricted. Like Charles and Walter, they have been educated not to value such perceptions. They are predominantly lawyers, business executives, and West Point soldiers, with an occasional actor thrown in. They are generally unfamiliar with art, history, and literature. Except in the sheltered manner of officials, they have rarely traveled in other countries. They would not be singled out for their refined sensitivity. Their chosen advisers are generally those whose sensory experiences have been confined to university libraries, classrooms, laboratories and hospitals, and suburban homes.

If you look at the expertise of the authorities in your world, you will probably notice a lack of any special claims to somatic authority. University presidents and deans, school principals, office managers, factory supervisors, and lab directors are rarely people we would single out as having felt or perceived more than the rest of us. They are more often those who succeed in pleasing those above them. One might object that such people have more experience than others in "managing," but until very recently managers have been trained in rational analysis, not in the perceptual abilities required to notice an employee's shifts of moods, an unheard complaint, an unrecognized tension. They have not been particularly skilled at helping people in their organizations work more creatively together. Some organizational theorists have recognized this sensual lack, and schools such as Yale and U.C.L.A. now include sensory awareness and encounter-group skills in their management training programs.

MOVING ON

Throughout this book I have associated "maintaining the status quo" with our physical training to adopt rigid stances, a "habit of performance." The muscular rigidities that come from holding those stances for years dull perception and make it difficult for people to adapt to each other and to new situations. But clinging to forms that worked for a few moments yesterday causes lower-back pain today.

In authoritarian ideologies, "change" is something extrinsic to the body. We are taught to fit into successive forms; different molds are stamped on us. Based on their sensual experiences, the somatic pioneers rejected that notion of change. Moshe Feldenkrais, the most articulate theorist of flexibility, writes of the status quo:

> I contend that rigidity, whether physical or mental, i.e. the adherence to a principle to the utter exclusion of its opposite, is contrary to the laws of life. For rigidity in man cannot be obtained without suppressing some activity for which he has the capacity. Thus, continuous and unreserved adherence to any principle, good or bad, means suppressing some function continuously. This suppression cannot be practiced with impunity for any length of time.[15]

Pause

At this moment your metabolism is provoking subtle adjustments in your body. Your breathing just shifted as you read this sentence. You may hear a new noise — a telephone ring, the sound of a passing truck, a child's cry. How does your body respond to this constant flow of stimuli? Do you try to keep yourself still? Do you ride the waves of change?

The body, like any organism, is a protean reality in constant flux. Our social institutions, however, fail to reflect this incessant movement. In dealing with children, for example, parents, physicians, and educators behave as if bodies were inert clay and require external force to shape them, rather than as biological organisms, fermenting *bottichs*. Forcing children into rigid shapes inhibits their learning capacities.

Putting orthopedic devices on two-year-old Johnny would have impeded his experiments in learning how to support himself on his legs. Sister Ruth's learning process was severely interrupted by surgery and pain-killing drugs. She was not treated as a person whose bodily pains gave meaningful signs about conflicts in her life, but as a corporeal machine whose parts needed adjustments. Feldenkrais's patient Nora was given drugs that inhibited her sensory awareness, which was her key to health.

Error comes from such impediments; it arises from re-
stricing people's capacities for adaptation.

You will recall that authoritarian ideologies have traditionally blamed error on the vagaries of nude sensation and
on unruly pulls of emotion. We need "objective" experts, they
tell us, to keep our proclivities from enticing us into illusion.
But if you examine your experiences of error, you may find
that they are often due to a lack of attention, a mechanical
response that prompts you to act without sufficient
experience. It is not that sensation is culpable compared to
the purity of abstract intelligence; sensation is ambiguous,
and requires refinement, experimentation, feedback, and correction.

I'm in a hurry to make an appointment. In my rush, I
crash into a car I failed to notice pulling out of a driveway.
I buy a new car which has a very different shape from my
former one. I don't yet have an accurate sense of its length
and width. As I go to park it, I scrape the car in front.

"Error" for the Greeks and "sin" for the Christians
came from the Greek word meaning to "miss the mark"
and from the Latin word meaning to go astray, to
wander off the path.

Think of Alexander's golf student trying to hit the ball
far down the middle of the fairway. If he flubs it again, he
can hurl his club into the woods and give up. Or he can keep
the same stance, try to keep his eyes glued to the ball, stub-
bornly swing the same way, and hope for some miraculous
change in the way things work. Or, noticing a tension in his
forearms, he might relax them. The next time, he hits the
ball a little more squarely. He tries a slightly different stance
and relaxes his shoulders. The ball sails even further out
there. Noticing with care the results of these adjustments,
he continues to make even finer adjustments, knowing that
he is on the way to learning a skill.

Several factors can evoke the golfer's faulty habit of performance. His mother might be on the sidelines telling him
not to strain too much. His father might be grumbling to him-

self that his son will never make it; he's not good enough. His coach might keep yelling "Keep your head down!" An old shoulder injury might restrict the fluidity he needs to swing the club accurately.

The harried golfer, habituated to swing improperly every time he takes club in hand, while his attention is sucked outward toward his teacher and the spectators, is one image of our society, locked into self-destructive patterns, disconnected from many of its own resourses. The challenge posed by the somatic pioneers is whether a radically different way of experiencing our bodies — as *bottichs* — can distill innovative ideas about how our stumbling social institutions might walk more surefootedly toward the goal we would like them to achieve.

You now have sampled distillates from a variety of *bottichs:*

1. the technology of authenticity, brewed from the experiences of its pioneers;

2. this book, distilled from my own bodily history;

3. the world views and problems generated by the sensual experiences of Walter, Charles, Sister Ruth, and Imelda; and

4. the connections among the many levels of your own experience, associated in this book with various "pauses." The question that remains is how to create a world in which these various liqueurs can be enjoyed rather than polluted or outlawed.

CHAPTER NINE

CONSENSUS

Up to this point I have emphasized what you can do as an individual to reconnect with sources of wisdom within your body. But individual work is limited. The more isolated a person is, the more impoverished will be the amount of sensual experience he or she has available for making decisions. The hoarse actor alone in his studio feels he is reciting in a perfectly natural way; the golfer whacking out balls on a driving range feels he is keeping his eye on the ball. Their viewpoints are limited by the number of stances they have been able to take. The mother taking care of her child alone in her home is cut off from the sources of knowledge women shared in more traditional communities. Truly reliable judgments are based on experiences that we have shared with others, honed by feedback, contrasts, and comparisons. Plato and Descartes were on the right path in pointing out the unreliability of individual perception, but they went astray in placing the blame on the nature of sensation rather than on individualism. To make significant alterations in the social molds that shape our perceptions — schools, churches, political organizations, businesses — we must work together effectively through a consensus about what to do.

Consensus comes from the Latin that means "a feeling or perceiving together." The technology of alienation makes consensus difficult to come by. Communities seem to be more pulled apart by divisive ideas than impelled by organic rhythms which might unite them. Recovery of our shared genius requires utilizing the somatic resources we share with animals for acting in concert.

Recall how a flock of geese moves through the air. They form a loosely structured *V* which undulates like a single kite responding to air currents. The bird at the apex holds its position for a brief time, then yields to another. As the geese approach a pond, they descend in harmonious succession. Trout dart in and out, under rocks and through cascades, each responding to particular stimuli but moving with the rhythms of the others. As a herd of antelope moves across a meadow, the animals' heads turn in concert at the sound of a branch snapping under a hiker's foot. Their white rumps swing in a unified dance as they move away from possible danger.

We experience moments like these in human intercourse. A dry conversation floods with feeling as I grasp how relevant what we're talking about is to a difficult personal question. Or the other person suddenly commands my presence by angrily objecting that I'm not listening to what she is saying. A brittle argument dissolves when someone begins to cry and reveals that a close friend is about to die. In a long discussion filled with conflict, a group suddenly happens upon consensus. Everyone nods together, and the discussion takes on a new life as people begin building with each other. In dancing and lovemaking there are moments when the intensity of pleasure overpowers the noise of our incessant inner dialogue. Awkward forms that make people move discordantly give way to rhythms that unite. When my hands conform to the muscles of a client, the flood of ideas that usually overwhelms my attention often dissolves in sensitivity to the other person's movements.

Norman O. Brown writes of this revolutionary shift of feeling:

> Union and unification is of bodies, not souls . . .
> soul, personality, and ego are what distinguish and

separate us; they make us individuals, arrived at by dividing till you can divide no more — atoms. But psychic individuals, separate, unfissionable on the inside, impenetrable on the outside, are, like physical atoms, an illusion . . . Souls, personalities, and egos are masks, spectres, concealing our unity as body. For it is as one biological species that mankind is one . . . so that to become conscious of ourselves as body is to become conscious of mankind as one.[1]

This was the kind of unity that Feldenkrais achieved with Nora as he gently moved her limbs and felt her hands. It is what I often saw emerging between Ida Rolf and the many children whose flesh she was sensitively manipulating. Each of the somatic pioneers articulated one or another aspect of how an individual body is related to the whole system. Performing in front of his mirrors, Alexander found that his laryngitis was not just a matter of germs infecting mucous membranes; it was also the result of the way in which he muscularly responded to expectant audiences. Wilhelm Reich uncovered the cultural sources of muscular armor as he tried to understand the relation between sexual impotence and the popularity of fascism. Feldenkrais and Rolf showed how the universal field of gravity affects one's life. Emilie Conrad found unifying organic patterns beneath the somatic ideals that different groups of humans impose on themselves.

When I feel that I am an atomistic self encased in a *corpus,* I exist in that world where people have fought for centuries over the meaning of god, freedom, justice, and truth, without coming to any effective planetary agreements. Some have succeeded in grasping the power to decide those issues, usually by force, occasionally by becoming the accepted experts. One group becomes unified by an idea, which separates them from other human beings who don't accept the idea. But to reconnect with my body as *bottich* is to realize that all living beings share the same air, are affected by gravity, exist within a common matrix of elec-

trical attraction and repulsion, and depend on similar processes of nourishment.

One can reasonably object that there is a more obvious
dimension to biological existence, what Thomas Hobbes
called the war of all creatures against each other. Species feed
off each other; animals devour their young; survival is won
at the cost of constant warfare. Nature does not look like a
benign consensus.

Such raucous aspects of the physical world are unde-
niably real, but the shape of our social body has been designed as if these were the only parts of nature. Instead of
capitalizing on the harmonious aspects of biological reality,
our social forms are built on, and enhance, our deepest fears.
We are habituated by those forms to think of each other as
separate egos, pitted against each other in a struggle for food,
land, friends, wealth, and sex. Fragmentations permeate our
entire society, enhancing divisions among races, classes,
neighborhoods, and professions. In the face of these fragmentations, the status quo retains its power.

You may have experienced at least a few frustrations in
being part of a group trying to organize itself for action
toward a single goal. Despite the fact that the group agrees
on its purpose — reducing the crime rate in the neighborhood,
providing better-quality education for children in the local
school, making working conditions more human, organizing
for disarmament — it will often lose its unified momentum
and fail to gain power to effect change. The fragmentation
has many dimensions: psychological conflicts, economic imbalances, breakdowns in communications skills, disruption
by outside groups, and so forth. But it also has a somatic
component. You might sense a brittleness in the atmosphere,
or notice that some people are posturing while others are
shrinking in discouragement or fear. Tensions build in the
shoulders and guts of some participants, alienating them
from others. You may notice, and even come to expect, that
some people will lapse readily into familiar postures, straining
to be right, pushing to convince, or gesticulating to win an
argument. Instead of being able to respond to each other and

the overall needs of the group, people react in mechanical ways, bouncing off each other and creating distance and hostility. Healing these social fractures is not just a mental but a somatic accomplishment. It requires sensory awakening and flexibility in our joints.

With a colleague, I recently conducted a workshop at a conference on "Psychology and Power," sponsored by a local organization dedicated to making the techniques of contemporary psychology available to the working classes. The people in the organization are committed to such goals as participatory democracy and personal intimacy. For two days, 350 participants, mostly professional therapists and social workers, sat for hours on plastic folding chairs arranged in rows in the customary manner, facing straight ahead to a raised podium where a succession of "experts" lectured in academic fashion, often reading from a script without looking at the audience. There was little time for discussion after each lecture. During the periods assigned for small groups, many people seemed to be so exhausted that they spent the time walking outside or chatting in the halls.

When the time came for our workshop, one of several given at the end of the conference, thirty-five bedraggled-looking people attended. The first thing I did was ask them to turn to each other in groups of two and discuss the various messages they had been getting from their bodies during the conference. The room suddenly came alive with excited conversation. I then asked them to recall with each other what their bodies had told them when they were in school, going back to their earliest memories. I noticed that several people began to talk angrily, mimicking various postures, gesturing freely. One woman, who appeared to be in her late fifties, said with tears in her eyes that she felt absolutely dead — she had felt that way all through school, and she couldn't remember having any bodily feelings during the conference. She said she felt furious about how passive she had been. Others spoke of old tensions and restlessness. Most were surprised to realize how passive they had felt during this meeting of political "activists."

I next asked these people to find ways of being in the room that enabled them to listen actively as my partner described a case study. I suggested that they experiment with different sitting postures, in chairs and on the floor, and give themselves permission to move about the room and stretch if they needed to. I encouraged them to let their gaze wander freely, and to look often at other people in the room as well as my partner, geting a sense of what was happening with all of us.

My partner, a Reichian psychotherapist, presented at some length the story of an anorexic young daughter of political activists, to illustrate what happens with people who have the right ideas about liberation but who are ignorant of the needs of the body. While she was speaking people in the group were livelier listeners than they had been during the conference: their eyes were brighter, their bodies more mobile. They began to show signs that they were getting a sense of shared authority for the group. When dullness came over the group as my partner spoke too obscurely, people took the initiative to interrupt and ask her to clarify what she was saying. It gradually came to seem as though we were all there exploring a problem together, rather than listening to an expert whose opinions we were either to accept or to pick apart.

This conference was a typical instance of a social institution that was "sick," in the sense that it was not using the available resources. Here were 350 people who had wide-ranging experiences in community organizations and rich intellectual and cultural backgrounds. The conference was organized to bypass their shared genius in favor of the expertise of a handful of people who were not even familiar with the local community.

The instance also illustrates how difficult it is to achieve real consensus in the usual atmosphere of physical immobility, which makes people actually unresponsive — deaf — to one another as well as to the speaker, leader, teacher, or consultant. What I did in the workshop was to give people a brief glimpse of how to share responsibility for the somatic aspects of the group: the arrangement of furniture, physical relationships to the speakers, ranges of movement, types of postures, kinds of physical interaction among the participants, and so on. When I work with people over a longer period of time, as with students in a university course or professionals in a training program, I aim to make the group habitually aware of its sensual resources, its shared somatic genius. That usually becomes an ordinary part of the group's atmosphere, freeing the people to focus more effectively on the task at hand. Sometimes a group needs specific techniques. If people seem restless when we're about to discuss a difficult book or listen to an intellectual presentation, we might do five minutes of quiet breathing. Late in the afternoon, when people may be sluggish, I might suggest a five-minute interlude of aerobic movements or stretches. I often invite people to touch each other on parts of their bodies where they feel the discomforts that often arise during a long day of work. Transferring the notion of *bottich* to the social body, I think of using these techniques to prepare for the refined brewing of ideas in our collective vat.

It is crucial to recognize the fine line that divides the technology of alienation and that of authenticity: the primary goal on one side is to make people more comfortable; on the other, aware of their shared genius. I could use the techniques to lull a group into a comfortable passivity so that people will listen to what I want to say. In that case I would use the technology to enhance my own authority and to make the group more responsive to that authority. This precise move transforms many otherwise useful techniques into methods for deepening people's sense of impotence. But, like Feldenkrais working with Nora, I can use the techniques to reconnect the group with its own sensibilities, so that the group can teach me how I might most effectively contribute to its goals.

Such an approach raises many challenging issues for experts like myself. As the "Psychology and Power" group became livelier, individuals spoke out more freely about what they wanted to happen, often without being acknowledged by my partner and myself. They didn't want to hear everything we had planned to say, and they had questions that we hadn't anticipated. We were forced to alter our picture of what we had intended to do. Similar experiences occur in my university classes: students want to read different books, change the syllabus, do unconventional term projects.

I sometimes become uncomfortable at such points. I am faced with the choice of either halting the consensual process, basing my claim on official authority, or letting myself ride the impulses that are coaxing us all toward a new status. "But surely I've been flown all the way from San Francisco to New York and paid a high fee because I'm supposed to give these people some *thing* that I have and they don't!" "I've been hired as a teacher because I know something that the students don't, and I'm supposed to give it to them or I'm not doing my job." The decision to opt for alienation or authenticity puts me up against my own sense of self-worth.

Think of my analysis of somatic ideals. What is problematic is not the idealistic notion that we all have more capacities than we utilize but our fixation on the visual. Training in the somatic ideals accustoms us to focus on the way things look.

Similarly, I can prepare for a lecture or a workshop by fixing on the way I want it to look. In that case, I will consider any interruption of that predetermined process of disruption, a distraction, that draws the group away from the course which I have designed. By contrast, I can allow members of the group to teach me how I can best contribute my unique skills to their learning process. In that instance I can't know in advance what the session will look like. I can prepare for it by gathering my resources on the subject at hand, but I cannot anticipate precisely what the group knows and what it would be like to explore until we begin moving together.

I suspect that at this point you can get a precise feeling for the challenges that the technology of authenticity poses for people in roles of authority. As long as people sit quietly in their rows of seats or desks and ask to be acknowledged by raising their hands, it doesn't matter how radical their ideas are. They are safely contained in a world alienated from this tangible room where I choose to stand in front and maintain my authority. Ideas are safe as long as they don't move people out of their chairs or into the hallways. They are even relatively safe if they impel one or two people to stand up unrecognized and yell out. I can say to myself, "There's always one in a crowd." But if the whole group begins to move, my attachment to the status quo is threatened.

Even if the group begins to move, however, its impetus may be short-lived. A genuine consensus requires a free interchange of information, a sharing of individual genius. Our sensual training, capitalizing on our primitive fears of survival, makes sharing difficult.

If I tell you everything I know about the body, you will not need me. I'll lose clients. You will steal my ideas, write books about them, and I'll fade into obscurity. People will start hiring you as a consultant instead of me. If I decide to name my work "Donning," and you learn all about it, you're liable to go out and teach it under your name and make money at my expense.

If a physician openly admits to a woman with a lump in her stomach how little authoritative information he has about it, she might go to someone else. He might feel less self-esteem — and if that happens frequently, his income might be reduced. If bankers, lawyers, real-estate brokers, corporate executives, office managers, politicans, and military experts openly shared their knowledge, they would have to confront radical changes in their psychological and economic attitudes. The technology of alienation unerringly shapes us to feel "It's either you or me, us or them." When I am in front of a group of individuals who feel that way, seated so that they can't see or touch one another, my own authority is not easily threatened, even by a rowdy or two. But when people turn to face each other, begin to notice each other's expressions, tensions, and emotional reactions, and even begin to touch one another, the group gains a strength that I cannot easily resist. I have to alter my own notion of authority or use defenses that may even escalate to violence.

In previous chapters I have emphasized how the technology of authenticity reconnects various layers within our personal experience: ideas, movement patterns, sensual experiences. In this chapter I am showing how that technology also reconnects us with one another on a somatic level, reestablishing links between my perceptions and yours. One may easily object that it is utopian to imagine that this technology could have any noticeable effects in the modern world, but there are several movements that indicate a growing receptivity to a consensual model based on our sources of biological unity rather than on our delusions of being separate egos.

TAKING HOLD OF OUR BODIES, OURSELVES

The area of health care is where the most dramatic examples of people learning to reclaim shared authority over their own bodies have occurred.

(1)

Many of the moves in this direction have been initiated by women dealing with the damaging results of their social isolation. One example is the Boston Women's Health Collective, which eventually published *Our Bodies, Ourselves.* Fourteen women, ranging in age from twenty-five to forty-one, had experienced similar feelings of anger and frustration in dealing with male medical experts. As they met with each other, they soon discovered how much they could learn about their bodies by gathering and evaluating medical information themselves. They looked into books and journals and consulted friends who were physicians, nurses, and medical students. Of particular significance to what I have written about "experts" was their realization of how much they knew about health care from their own shared experiences. This collective data bank became richer as they gave classes for other women. They write of the results of their collaboration:

> We are better prepared to evaluate the institutions that are supposed to meet our health needs . . . For some of us it was the first time we had looked critically, and with strength, at the existing institutions serving us. The experience of learning just how little control we had over our lives and bodies, the coming together out of isolation to learn from each other in order to define what we needed, and the experience of supporting one another in demanding the changes that grew out of our developing critique — all were crucial and formative political experiences for us. We have felt our potential power as a force for political and social change . . . For us, body education is core education. Our bodies are the physical bases from which we move out into the world; ignorance, uncertainty — even, at worst, shame — about our physical selves create in us an alienation from ourselves that keeps us from being the whole people that we could be.[2]

That quote sums up everything I've said in this book. In isolation, the women sensed little control over their lives and bodies. Body education is core education: ignorance, uncertainty, and shame about our physical selves alienates us from ourselves and keeps us from being the whole people we could be. Learning from one another to define and get what we need for our bodies is an act with significant political implications.

Our Bodies, Ourselves, which has inspired similar books for both men and women, presents simply and graphically a wealth of essential information about the body and relates that information to problems arising in menstruation, pregnancy, contraception, abortion, and common illnesses. The book is written to enable the reader to gain authority over her own health care and to learn how to evaluate the advice given her by others. It is reminiscent of the *Domestic Medicine* books popular in the nineteenth century, which I cited in my second chapter.

Similar groups of women have come into being throughout the country. Some have been inspired by the Boston collective, and others have been initiated by feminists, social
workers, and nurses. Nurses, for example, have gone into
working-class neighborhoods of several cities and trained
women to take care of themselves and their families. One
nurse, Dolores Krieger, has initiated a nationwide network
of classes to make people aware of the healing power of the
kind of sensitive touch that nurses apply to their patients.[3]
Nurses in Third World countries have taught people in local
communities the skills necessary for everyday health care.[4]

(2)

Several physicians have initiated changes in the popular
model of medical authority. For example, Dr. Tom Ferguson
began a movement similar to the Boston Women's Health
Collective's. While a medical student at Yale, he realized that
the vast majority of people go to doctors with complaints
they could resolve themselves if they had the most basic
information. In 1977 Ferguson published the first issue of
Medical Self-Care, a journal with the purpose of dis-
seminating useful data for taking care of one's health and
for evaluating the opinions of various medical experts.

Also during the past decade several physicians have
created clinics that bring together healers from various traditions: somatic therapists, acupuncturists, herbalists, bio-
feedback technicians, and psychologists. These clinics
represent a radical departure from the prevailing system of
health care, in that they are based on an educational model,
training the patient to take charge of his or her own health.
A person coming to one of the clinics is given the opportunity
to be diagnosed by practitioners of several ways of healing
and is also trained in techniques of sensory awareness,
visualization, and relaxation. This training puts the person
in a better position to evaluate which diagnosis will be the
most useful.

In Chapter Two I cited Dr. Eva Salber, a professor of

family and community medicine who has taught health professionals throughout the world how to seek out and use what she calls the "natural health facilitators" within local communities. I also have mentioned the increasingly large numbers of parents who are assuming authority over pregnancy and childbirth. You may have noticed that the image of health care produced by these movements is not unlike that of the decentralized health-care system of pre–twentieth-century America, offering a diversity of healing methods and stressing the need for people to educate themselves about their bodies. Such an image is more in keeping with the democratic proclivities of early Americans who resisted importing foreign hierarchical models of authority.

In addition to these and many other health movements are groups whose goal is reversing the perverted system of social values that puts abstract ideologies ahead of fleshly human lives.

(3)

Food First was begun nearly ten years ago by Frances Moore Lappé, author of *Diet for a Small Planet,* and Joseph Collins. Now numbering some 15,000 members spread throughout the world, this organization and many like it are engaged in teaching people how to reclaim their ability to grow their own food, food that is truly nourishing, from the extensive agribusinesses. Food First has exposed the manner in which large corporations have depleted the soil by excessive use of fertilizers and polluted food with pesticides, robbing local communities of the ability to care for their own needs. Members of the group are teaching people in several countries basic information about cultivating the soil and growing crops.

(4)

Several peace organizations reflect a consensual vision of social institutions. Physicians for Social Responsibility resulted when several physicians woke up to the absurdity of devoting their lives to health within an ideology that is gearing up for mass destruction. They have mobilized thousands of physicians throughout the world to protest the endangering of bodily life for the sake of tenuous ideologies. Their activity has spurred other professionals to create similar organizations: Educators, Psychologists, Artists, and so on . . . for Social Responsibility.

The Mo-Tzu peace teams, named after a fifth-century B.C. Chinese peacemaker, travel to various hot-spots throughout the world — Israel, Egypt, Northern Ireland — establishing contacts with "natural peacemakers" in local communities. Like Eva Salber in her search for "natural healers," these teams hope to give added strength to local leaders. One of the members speaks of the sensual dimensions of this venture:

> At some point it becomes dangerous to theorize about war and peace from the comfort of our living rooms. Dangerous because we risk losing touch with reality. In the middle of these conflicts, in the middle of these landscapes, when we are surrounded by the smells and colors and shapes of these people so different from ourselves, thinking itself becomes different, as do the thoughts. We discovered again the need to open ourselves, to let in the strange and alien feelings, to identify, if only for the moment, with an Israeli's grief for a fallen son or a Palestinian's grief for a lost homeland. Only by joining, at least for the moment, in their feelings could we hope to speak back into these people's lives.[5]

(5)

A group of California politicians, under the leadership of a senior member of the State Assembly, John Vasconcellos, have created an informal network devoted to passing legislation that supports fundamental biological values: humane births, healthy sexuality, educational policies that respect the bodies of young students, and enlightened food policies. Vasconcellos has written of this movement in his book, *A Liberating Vision: Politics for Growing Humans.*[6]

(6)

In the previous chapter I referred to recent findings that old hierarchical forms are failing to keep companies moving with sufficient flexibility, and there are some signs that the business world is moving toward more consensual models of organization. For some two decades places such as the National Training Lab in Maine, the Menninger Foundation in Kansas, and the Western Training Lab in California have trained managers in sensory awareness and group decision-making skills. Organizational psychologists such as Will Schutz, Robert Tannenbaum, and Herb Shepherd have argued that a more decentralized consensual form of management that capitalizes on workers' shared genius would enhance productivity.[7]

Many observers of the current economic scene point to signs that the efficacy of traditional approaches to business is rapidly diminishing. The world monetary crisis, the tenuous state of our economy with its record number of business failures, rising worldwide unemployment, uncertainties about energy, and increasing incursions by Europeans and Asians into markets historically dominated by the United States are evidence of the need for radical changes. Moreover, the shift of emphasis from heavy industries toward service industries and electronics requires profound shifts in the organization of the workplace.

The futurist Paul Hawken argues that we have passed beyond economic eras of accretion (imperialism and colonialism) and replication (mass production) into an era demanding models based on mutuality.

> The continuous extraction of resources which are blown into the sky through smokestacks to make plastic paper towel holders no longer satisfies. The whole biosocial

organism of our economy coughs and has fits. The dispersion of the raw economic power towards the margins creates a de facto type of political system in which social contact and mutual respect are a far stronger cement than money and wealth.[8]

I've left for last the institution most important to the shaping of our attitudes toward authority, and the most resistant to change: public education.

Educational theorists such as Rudolf Steiner, Maria Montessori, and John Dewey have argued for decades that a sound education must be grounded in the development of a child's sensibilities through bodily expression and the arts, and Sir Herbert Read wrote prophetically in 1959: "Unless we discover a method of basing education on these primary biological processes, not only shall we fail to create a society united in love; we shall continue to sink deeper into insanity, mass neuroses and war."[9] Despite these numerous voices calling for reform, schools have changed little in their policies toward the bodies of students. (Preschools, some private schools, and alternative schools are the exceptions.) Nonetheless, I feel that it would be unfair for me to conclude this chapter on the social implications of the technology of authenticity without at least imagining how it might be used in our schools.

Let's say I've been hired as a consultant by an elementary school. The first thing I would do is set up a series of intensive weekend workshops with the teachers who have something to do with somatic education: coaches, physical-education instructors, and teachers of art, music, dance, and theater. I would educate them in basic techniques of sensory awareness and body movement, giving them ample opportunity to discuss with each other memories of what it was like for their own bodies when they were students. I would enlist the help of more expert colleagues to give them some experience of how sports and dance can be taught in the non-authoritarian, functional ways I've described in this book.

Next I would ask these teachers to join me in designing

a series of perhaps three intensive weekend workshops for all the teachers, administrators, students, and their parents, to communicate this somatic approach. A result of these workshops would be to create guidelines for a long period of experimentation in redesigning the somatic environment of the school. I would then train the somatic educators to work with me as consultants in individual classes, helping the teachers use the new technology in achieving their particular class goals. That might include setting up the classroom so that students could move about quietly and have various options for changing their positions. Teachers might want to provide cushions as well as desks and different kinds of writing surfaces. Students might be encouraged to do quiet stretching exercises when they become tense or restless.

We would examine the daily schedule from a somatic perspective. One class might decide to begin each day with ten minutes of stretching exercises followed by five minutes of quiet meditation. In mid-morning they might want to have fifteen minutes of vigorous movement. Perhaps they would schedule three or four five-minute periods of relaxation exercises throughout the day. Physical education would not just be sandwiched in where it is convenient. At the present time, for example, students are often put in P.E. classes immediately after lunch, a period that is so short that they rarely have time to shower and cool down for their next class. We would have to experiment to find the most appropriate hour and length of time for physical education.

Application of the new technology could be a major step toward alleviating the discipline problems that now consume an enormous amount of a school's energy. Imagine (or remember) what it is like for young children with rapidly changing bodies to have to sit still for several hours in hard desks memorizing arithmetic tables, historical dates, and geographical facts. Or what it is like for young teenagers filled with newly felt sexual juices to be forced to stay quiet while teachers drone on about algebraic factoring. The teachers themselves don't do too well, often complaining about chronic weariness and strain. Techniques of focusing on breath, muscle tension, or fantasies might help the students to quiet themselves for concentration on intellectual work. Both students and teachers would profit from learning how to relieve tense muscles, headaches, and eye strain. Vigorous movement at certain times might ease pent-up energy.

Some people might claim that this would take so much time that there would be no time left for traditional learning. My guess is that these methods would rapidly accelerate the learning process. Getting rid of the muscular and emotional barriers to being in school would free the teachers to become educators rather than disciplinarians. So much time is wasted in schools just trying to get students' attention.

The deepest barrier to applying these methods is the current school system's roots in the mind-body split. The more "mental" the discipline, the more value the school system gives it — math and science at the top, physical education at the bottom. My consulting efforts would aim to teach the significance of sensual education in relation to other aspects of learning. This could require more sophisticated training for those I've called the somatic instructors.

Sports, for example, would look very different. They are presently taught according to an idealistic authoritarian model: "Place your hands on the bat this way and keep your eyes on the ball"; "When you approach the hurdle, place your right leg here and lift your left leg like this;" "You dribble that ball like a sloppy fish, why can't you do it right?" Many people, like myself, may remember the discomfort of feeling clumsy, foolish, or awkward for dropping an easy fly ball or missing a shot into the basket. Students learn to associate sports with shame about first pubic hairs, budding breasts, narrow shoulders, and clumsiness.

In my fantasy, physical education in the early years would emphasize sensory awareness rather than winning. Teachers would help students experience the pleasures of breathing rapidly, feeling blood course through their legs, moving together in harmony with the group, feeling the strength of their arms and the range of movement in their joints. Skill would come slowly as students learned to discover by themselves the muscular requirements demanded by the weight and shape of a basketball, a bat, or a hockey stick. They would be given lots of time to find out by trial and error what stances and movements work best for them. Teachers would introduce the rules of any particular game slowly, always retaining the emphasis on awareness and experimentation.[10]

Physical education classes are an ideal place to give young people the chance to explore the relations between muscles and emotions. A trained somatic teacher can use experiences of anger, jealousy, and shame to give students a deeper sense of themselves and each other.

I would also broaden the notion of somatic teachers to include teachers of art, music, and drama, since these subjects are fundamental in learning about the body. Music can teach how a player's entire body affects the sounds coming from an instrument. Mime and ensemble theater teach students to be sensitive to the most subtle movements of their own and others' bodies, and drawing and painting give students an opportunity to explore the relation between the eye and hand and the whole body.

In my vision, young children would slowly learn to brew abstract ideas from their various somatic experiences. For example, each student might draw maps of his or her felt body, outlining its various parts, channels of energy, relative sizes, and so on. Children could compare these maps to each other's and to traditional maps — Western anatomical, Vedic, Chinese. This process would provide an enormous amount of information about various cultures, religions, and systems of social organization. Students could be led into the more abstract kinds of sensory pattern recognition that characterized the historical beginnings of mathematics and geometry. Teachers could use the students' refined somatic experiences as an experiential basis for teaching the physical concepts of gravity, inertia, momentum, and resistance. Various social studies might be more vivid if they were anchored in increased somatic awareness. Instructors could root anthropology and sociology in an understanding of different uses of the relationship between the body and power as manifested in slavery, war, religion, medicine, and agriculture.

As you can see by the few suggestions I've made, such schools would involve a regular flow of creative suggestions from the teachers. They would not be like Summerhill, where students were left to their own private devices. I do not equate "permissive" with "flexible." I do envision schools in which the students and teachers have freedom to move around and adopt more energetic postures rather than being rigidly confined to desks in rows. I anticipate that students and teachers would cooperate to design their environment to promote bodily vitality during class. The emphasis would not be on each student "doing his own thing" but on consensus, on shared somatic states in which there is constant interaction among all members of the school to create an exciting learning environment.

Moreover, I'm not asking teachers and administrators to give up their authority, but to discover a deeper source of personal authority and expertise. In the present system they are constantly being ridiculed behind their backs and irritated by the various disruptive strategies of their students. My fantasy would allow teachers really to use their skills and their experience — the basis for genuine authority.

In Chapter Four I sketched the case of two-year-old Johnny, whose doctor prescribed braces for his struggling legs. The pain of our social world comes from constantly being forced into braces of one sort or another, designed according to the bright ideas of some expert. Those devices inhibit our movement and obstruct our sense of how to learn. As children we cry aloud or get sick in response to the discomfort provoked by such devices, but as we grow older we learn to internalize our pain and transform it into quiet despair or psychosomatic illnesses. The challenge posed by somatic pioneers such as Elsa Gindler and Moshe Feldenkrais is whether we have the courage to try walking without those braces and to help one another in our halting efforts to find a new sense of balance.

CHAPTER TEN

CONSENSUAL SPIRITUALITY

While drowsily awakening from my dream about orthopedists, which I described in the first chapter, I first associated it with the resolution of what I would call a spiritual conflict. I have alluded to that conflict when sketching the history of the two sources of our word *body*. The Mediterranean and Sanskrit traditions defined a person composed of two parts: a soul, which belonged to the spiritual realm governed by God and peopled with angels and devils, and a *corpus,* a disconnected material object governed by the laws of physics. Spiritual authority came from above. Access to truth was through sacred books, their official interpreters, and those few people believed to have direct access to that other world — popes, evangelical preachers, psychics, rimpoches, and gurus. The Northern European tradition defined a person as a single reality, a body, a *bottich.* That definition implied a spirituality distilled from organic processes, a refinement of perception, emotion, and worldly action. Spirit was not a thing belonging to another world but a dimension of this palpable world. It arose in those moments of love, insight, and shared trance states which gave people the sense that they could transcend the apparently thinglike, impenetrable aspects of matter. Spirituality meant liberation.

Like the dispute in my dream, the contrasts between the two traditions often surfaced in arguments about the

196

nature of authority and sometimes became openly violent. The corporeal tradition is reflected in theologies that give certain individuals ultimate authority over their disciples. Their power is thought to be derived from divine commission or privileged mystical illumination. The *bottich* tradition roots authority in the innate genius of the people (the "enthusiasts"); other kinds of authority are derived from their consent.

The dialectic between *corpus* and *bottich* was engraved in my flesh by a religious education that mixed earthy druidic folklore with the mystical fantasies of Byzantium. My mother and my first religion teacher, Sister Felicia, nourished my early sense of a world more marvelous than what I saw around me. They said that God's Holy Spirit moved within me, always guiding and protecting me from harm. Prayer, they said, was listening to those impulses. If I always paid attention, when I died I would go to heaven — a bodily place where I would have all the games I could ever want, delicious food, the best toys, and playmates who would always be nice.

During my early years in elementary school, the physical side of Catholicism relieved the boredom of everyday life in the Sacramento Valley. I felt spiritually refreshed by the drops of holy water sprinkled on me at high mass on hot summer Sundays. Marching in processions, holding candles, and singing rousing hymns accompanied by organ and kettledrums were a respite from my feelings of isolation; I was part of God's enormous horde of saints. I sometimes got tears in my eyes when the priest rubbed ashes on my forehead and said I was made from dust and would eventually return there. When I reached the "age of reason" (which theologians in 1940 set at about eight years old; this was pushed back considerably by Vatican II), I was allowed to receive Holy Communion, which Sister Felicia described as eating the flesh and blood of Christ in fact, not symbolically as the Protestants said. I became an altar boy and could participate directly in rituals in which scores of priests, deacons, and acolytes moved according to fifteen-hundred-year-old choreographies, garbed in gilded silk brocades and speaking an ancient sacred language that made God visible and audible. I became intoxicated by clouds of incense as the choir intoned medieval chants. We stood or knelt in a sanctuary mystically illuminated by the sun filtered through stained glass windows that depicted Adam and Eve delighting in Eden, Melchisedek giving Abraham bread and wine before the gates of Jerusalem, Jesus sweating blood in the Garden of Olives, and Francis of Assisi in mystical rapture on Mount Alverna. I sensed myself to be a part of a very tangible history that reached back into the creation of the cosmos and extended forward into an eternity in which we would embrace each other in a garden of earthly delights.

Father John, my catechism teacher, told me that Protestants thought God and heaven were in a world apart from this, known only by faith and not by reason or sensation. We believe, he said, that God actually touches us in the holy water, the eucharistic wafer, and the healing sacramental oils. We see God in the ritual movements of the priest and people; we hear him in the Gregorian chant and the words of the mass. God revealed this world's meaning to us by making himself into a body in the person of Jesus, who had perceptions and emotions just like we do, with some minor exceptions: he didn't get erections or feel resentful toward others, like I did. It made a lot of difference, said Father John, what you did with matter. You had to use just the right kind of wine (21 percent alcohol), make ritual gestures just as they had been done for two thousand years, and wear the colors appropriate for each festival (black for funerals and All Soul's Day; red for Pentecost). An altar boy who manipulated the crystal cruets sloppily was being spiritually sloppy.

While I was learning the spiritual significance of matter and the importance of listening to my deepest impulses, Sister Felicia and Father John were teaching me a seemingly contradictory viewpoint. They said that Satan was lurking deep inside me, fighting with God to lure me into activities that would earn me eternal damnation. Satan was history's cleverest deceiver; he could make me think he was God. He knew me so well that he could easily draw me toward evil without my catching on until I was hopelessly addicted to his pleasures. You must be constantly watchful, they told me, over your desires and impulses. The moment you are inattentive Satan will leap in and take control. And you're doing God's will. The only way to be safe is to obey the laws of the Church to the letter and do what the priests and nuns tell you. Otherwise you will be deceived.

Even before I began to experience the genital urges of puberty I felt inwardly torn apart. I was supposed to find God within, but I couldn't trust what I found inside me because it might be the devil, master of deceit, dressed like the angel he once had been. I constructed rigid muscular barriers against the onslaught of diabolic impulses from within and turned my loyalties to external authorities.

Brother Philip, my high-school religion teacher, raised questions about those authorities which made me even more unsure about the path to reliable knowledge. Jesus came to free people, according to Brother Philip. Serving God didn't mean obeying a lot of external rules like not eating meat on Friday, going to mass every Sunday, and keeping your hands off your cock — that's what the Pharisees did, and Jesus called them hypocrites. Serving God means getting rid of the baggage that keeps you from being free and loving — your attachments to a cozy life, your fears that people won't like you. Christianity means risking your life to liberate the world. You can't do this, Philip said, unless you have the courage to follow your deepest convictions.

When I entered the Jesuits at twenty-two years of age, motivated by Brother Philip's radicalism and my desire to loosen the devil's hold on my flesh, I found even more extreme paradoxes. Father Healy, my master of novices, based his spiritual direction on "The Rules for the Discernment of Spirits," written by Saint Ignatius Loyola, founder of the Jesuits. According to Ignatius's teaching, you made decisions about how best to serve God by putting yourself into a quiet meditative atmosphere for several days or weeks and listening to the various impulses that moved you as you considered various options. A spiritual director could assist

you in interpreting these impulses, somewhat in the manner
of a psychotherapist, but could not relieve you of the ultimate
responsibility to decide which impulses revealed God's will.
The methods the Jesuits used were similar to the methods
developed by Charlotte Selver and Emilie Conrad. We
meditated with different techniques such as guided fantasy,
breathing, and mantralike repetitions. Often we simply sat,
quietly allowing the flow of thoughts and images to pass by
without censorship. We varied our postures, diet, times of
meditation, amount of light in the room, and types of reading.
In all these experiments we were taught to observe the
changes in our inner impulses and try to discern a pattern
that would indicate the Holy Spirit's personal directions.

Father Healy placed what I then thought was exaggerated emphasis on the body. His daily lectures were sur-
realistic collages of readings from books on psychosomatic
medicine and mystical literature. His fundamental text for
integrating these collages was the Jesuit Herbert Thurston's
The Physical Phenomenon of Mysticism.[1] Thurston argued
that the intimate bond between the divine and the body
implied that prayer and the sacraments should have a direct
impact on the flesh. His research into the history of
mysticism confirmed this thesis. He described bodily
phenomena regularly reported by mystics: levitation, tele-
kinesis, luminosity, immunity from being burned by fire,
living without food, and emitting fragrant odors. Presaging
my future interest in Ida Rolf's techniques for lengthening
bodies, I was particularly taken by the phenomenon of
elongation as reported, for example, in the canonization
hearings of Sister Veronica Laparelli:

> On one occasion among others, when she was in a
> trance state, she was observed gradually to stretch out
> until the length of her throat seemed to be out of all
> proportion in such a way that she was altogether much
> taller than usual. We, noticing the strange occurrence,
> looked to see if she was raised from the ground, but this,
> so far as our eyes could tell, was not the case. So, to

make sure, we took out a yard measure and measured her height, and afterwards when she had come to herself we measured her again, and she was at least ten inches or more shorter.[2]

Father Healy (now a Jesuit phrenologist) said that if we adhered to the inner impulses of the Holy Spirit and were faithful to meditation, watched our diet, and exercised regularly, our bodies too would be transformed in unsuspected ways. He emphasized that the daily eating of Jesus' body in the eucharist was transforming our flesh into immortal flesh.

Paradox was rapidly changing into conflict. On the one hand I was supposed to regard only the authority of the Holy Spirit, operating within my muscles, nerves, and guts. On the other I had taken vows of poverty, chastity, and obedience which drew me totally outside myself. Father Healy said that the vow of obedience required us not simply to obey whatever authorities told us, "like an old man's staff," but to force our thoughts in line with theirs even if they said something was black that we saw as white. The vow of chastity enjoined not simply abstention from sexual intercourse but acting, Saint Ignatius wrote, "like the angels" — refraining from anything the least bit redolent of sex, like slapping a fellow Jesuit on the back during a baseball game or hugging our parents. We were also to keep careful watch over our movements and our senses. Father Healy told us to keep our eyes cast down when we walked about and deliberately not to listen to noises that might distract our inner calm. We were to walk with measured paces, never running except to escape fire and the occasional earthquakes that rocked our monastery situated on the San Andreas Fault.

In my study of theology I found that my puzzles about where to find reliable spiritual truth — within or without, in the body's impulses or in abstract directives — were mirrored in recurrent rebellions against official authorities by subgroups in every religious tradition. Some have called these people "enthusiasts" (from the Greek, "infused with the spirit") because they have consistently argued that external authorities derive their power from the consent of the enspirited community.[3] The Christian tradition includes the second-century Montanists, the medieval *fraticelli* who inspired Luther's revolt against Roman imperialism, the so-called "charismatic churches," and modern Third World sects motivated by "liberation theology." These subgroups are often composed of people on the fringes of society who worship together "in the Spirit," speak in tongues, heal by the laying on of hands, and are often persecuted for refusing to submit to official authorities. They base their practices on descriptions of the early Church in Paul's letters. Their notion of communal authority is derived from the words of Jeremiah, quoted by Jesus to the Samaritan woman at the well, that the coming of the Messiah means the end of patriarchy. When the Holy Spirit is poured into the hearts of all believers, there is no longer need for any external spiritual authority. All share equal access to ultimate truth.[4] These subgroups also share the belief that heaven is not another world but the apocalyptic transformation of this perceptible world.

I found similar countercultures in other parts of the globe: Sufism and Bahai in the Middle East, Tantrism in India, and Taoism in China. They shared with radical Christians a rejection of external authorities as the ultimate source of truth, and a belief that *this* world constitutes the ultimate reality.

In addition to these conflicts about the nature of authority I discovered a long history of conflict about the significance of the human body. Father Healy's radically physical spirituality represented mainstream orthodoxy. Early ecumenical councils vigorously, sometimes raucously, debated about the nature of Jesus' body: did he really sweat, urinate, ejaculate, and feel pain like the rest of us, or was it all a divine charade? Some theologians with Platonic and Gnostic proclivities argued for the charade theory on the assumption that the material world is totally separate from the divine. God is elsewhere, they argued, glimpsed only through the eye of the soul. But Church consensus was consistently in favor of the reality of Jesus' flesh and the divinity of matter.

At the same time that they were affirming the divinity of matter, however, the Church fathers acted as if they considered it diabolical. Origen cut off his balls because he was inspired by Saint Paul's statement that it's better to be a eunuch than to burn in hell. Saint Augustine's unhappy love affairs led him to erect an antisexual theology that shaped generations of guilt-ridden European Christians. "Post coitum, omne animal triste est," he wrote; "After intercourse, every beast is unhappy." Monks such as Saint Anthony, who fled to the desert hoping to escape urban temptations, overcame their incessant cravings for sex and power only by beating and starving their bodies.

The early spiritual experts wove elaborate theologies to explain the conflicting teachings I had learned as a child: God manifests himself within your deepest impulses; so does Satan. Your own experience, therefore, is essentially unreliable. Being guided by your impulses can lead you to heaven or hell. (We knew at this point in history that personal impulses could produce a Hitler as well as a Saint Francis. And in my later years, some smugly pointed out that Charles Manson and Jim Jones had followed the promptings of their inner experience.) The only way you can be sure of escaping delusion is to rely on outer authorities such as sacred Scripture and the people officially sanctioned to interpret its ambiguous messages.

During the 1960s, while I continued my practice of Jesuit spiritual techniques and study of theology, I had several experiences that made me appreciate for the first time Father Healy's and Herbert Thurston's stress on the relation between bodily transformation and spirituality. In encounter groups with people like Carl Rogers I became aware of the anger and fear moving within my body. Training in sensory awareness with pupils of Charlotte Selver gave me a sophisticated sense of my muscles, breath, and feelings. Psychedelic drugs, the practice of hatha yoga, and oriental spiritual techniques enhanced that body awareness. I went to Esalen, the founder of which, Michael Murphy, had been inspired by Herbert Thurston's book as well as by Sri Aurobindo's emphasis on the relation between transforming the body and spiritual awakening. It seemed odd to me at the time that such worldly techniques could contribute to growth in spiritual awareness — how sitting in a Zen meditation room counting breaths, for example, could lead to the kinds of transcendent experiences that I hadn't had since I was a child enchanted by Father John's rituals.

Sister Ruth's experience was similar to mine. Since she was a little girl, she remembers, she always had an abiding sense of God's presence. Throughout her life she felt little connection between that sense and official religious activities. "As a kid, I hated mass on Sundays — ugh!" she says. "A priest way up on the altar droning on, paying no attention to me." Seeking a fuller experience of God's presence led Sister Ruth into the convent. "But it was the same thing. I spent hundreds of hours praying the way Jesuit priests told me, but I just felt pain in my back, loneliness, and boredom."

Within a few weeks of taking up racquetball and sensory awareness, Sister Ruth discovered that feeling for God erupting into her prayer life. During her morning meditations it emerged from subtle pulses in her blood and breath. Her work with Emilie Conrad's "Continuum" enhanced that awareness. She studied with the cultural anthropologist Joan Halifax, who taught her how American Indians use chant to intensify such experiences. Having long known about the charismatic movement among Catholics, Sister Ruth now sought out its practitioners and joined in their rituals, discovering that she was easily able to speak in tongues and perform healings with her hands. She eventually became the director of a retreat and conference center, where she educates Catholics in using these somatic techniques to increase their awareness of God's presence.

The technology of authenticity affected Sister Ruth's regard for authorities. She began to express resentment about her years under a male-dominated hierarchy when she realized that their medical and spiritual prescriptions had been of little use to her. She felt frustration about her role in an organization in which she could have so little authority because she was a woman.

Similarly, my new respect for my feelings and perceptions gave me a more critical look at established authorities. Jesuit superiors and popes often seemed more interested in worldly security than in Jesus' message of love and truth. Following official spiritual directives began to seem at odds with living a truly spiritual life. At the same time, my study of Vietnam upset my naive faith in the United States government. I left the Jesuits, turned in my draft card, and embarked on a ten-year healing journey into the realm of glands and collagen fibers.

My orthopedist dream tells how I have resolved the conflicts I used to feel at this stage of my life. The orthopedist in the white smock who prescribed the cast felt to me like teachings about hell, sin, the unreliability of sensuality, and the need to conform to ideal bodies. The longer-haired orthopedist represented more vital and liberating teachings: Sister Felicia's notion of the Holy Spirit pouring into our bodies, Brother Philip's revolutionary Jesus, Veronica Laparelli's gain of ten inches in prayer, Charlotte Selver's sensory awakening. My response to the doctors' advice in the dream reflected my knowledge that the proper way to heal my fracture could not be defined by anyone outside myself, no matter how enlightened he or she might be. At the same time I realized that it would be harder to heal my leg by myself in total isolation. I needed dialogue with friends who loved me and would share their experiences with me, challenging my own limited viewpoints.

This dream is about what I call "consensual spirituality." It is *sensual* in that I conceive spirituality as a refined distillate of sensual experience. It involves exploring the furthest reaches of perception in meditation, trance, ritual, fantasy, and fasting. It is *con*sensual because spiritual techniques help get rid of the barriers that keep us from being in love, those egoistic rigidities that prevent us from serving other people. The rituals that delighted me in my youth and the similar rituals that characterize all spirtual traditions produce the somatic states in which people can experience what Norman O. Brown meant when he said, "Union and unification is of bodies, not souls . . ."[5] Bodies not as *corpora* but as *bottichs,* roiling and exuding fragrant odors.

The conversation among the three of us in the dream suggests another dimension of consensus: reliable spiritual knowledge depends on dialogue. The writings of the spiritual pioneers consistently testify to the dangers of isolation. For one who has an active sense of the divine, it becomes easy to mistake the self for the divine, charged with the mission of directing the lives of those less favored and admitting no peers. The history of religion has several major chapters on the fanaticism that thrives on the isolation of a leader from his or her equals and on the fragmentation of the dependent disciples. What often goes under the name of "consensus" — say at Jonestown or Tehran — is actually the most extreme form of people's abdication of sensual authority to some Svengali who admits no dialogue.

I no longer seek spiritual masters or mistresses. I do, however, keep my eye out for friends along the way who will be honest with me, challenge my tendencies to be self-righteous, help me to clarify my uncertainties about the lay of the land, and appreciate how far I've come. Only in an atmosphere of love, honesty, and experimentation can I wend my way through the labyrinth of neurons and intestines without ending in confusion or, in the worst fantasy, madness.

For me, such a spirituality is more a quest than a reality. The only place I've found anything like it is among the Indians of the southwestern United States: the Hopi, Zuni, and Pueblo Indians of the Rio Grande Basin. But their spirituality is a distillate aged for centuries in high deserts, far removed from the centers of so-called Western civilization. My spirituality has to be distilled from Viking and Celtic grains, fermenting in waters from the Sacramento delta and laced with Mediterranean spices.

DEATH

I would be evading an obvious question if I were to close this book without a word on death. It is the principal issue for any spirituality. Fear of it leads to the creation of ideologies and ideal bodies. It changes human faces into the face of the enemy. Sexism is rooted in it. People adopt dualistic spiritualities to reduce its terror.

Within a corporeally oriented society, people can take two basic stances toward death. For the spiritually minded, death means the dissolution of the bond that chains the soul to a material world of no ultimate significance — the liberation of the self from its grimy prison. For the scientific humanist, however, death is simply the end of a personal history; reality is nothing more than particles moving in space, taking on different configurations.

My experience of a friend's death gave me a glimpse of an alternative to these views of death which seemed more in keeping with thinking of a person as a *bottich*. I met Elsie in 1975, when she was seventy years old. She and her husband, who had died ten years earlier, had come from Boston to New Mexico in 1939 and homesteaded one hundred acres in a barren strip of badlands between Sante Fe and Taos, at the base of the Sangre de Cristo Mountains. Elsie was an artist; her husband had been an engineer.

A physician had sent Elsie to me for Rolfing to relieve the tensions from what seemed to be a malignant tumor in her abdomen. She had refused surgery and chemotherapy, having lived her life, she said, as close to nature as possible. She was then running her homestead with the help of a twenty-one-year-old man named John. The two of them took care of an orchard of one hundred apple trees and a vegetable garden large enough to feed several families. She spent hours outside working every day. She also made puppets and gave performances during the summers at a local dude ranch. She was what I think of as a typical Yankee: irreverent, tough, cynical about pretentious authorities, loyal to her friends, and filled with good humor.

During the two years I knew her the tumor grew slowly, but Elsie complained of no special discomfort. We talked a lot about death. She had been with both her mother and her husband when they had died of cancer. She was afraid.

Her condition changed rapidly in the last week of February 1977. She started losing strength and felt a lot of pain. Her physician organized thirty of her friends to take turns with her at the homestead around the clock. She refused to go to the hospital and would take only mild pain-killers. At five o'clock in the evening of the fifth day, a friend telephoned to say that Elsie was fading rapidly and had asked for me. I drove out along the Rio Grande Valley. The temperature was about twenty degrees; the sky was perfectly clear. As I drove east along the dirt road into her place, the full moon was just rising over the snow-covered peaks of the mountains, red in the sunset like the blood of Christ after which Spanish settlers had named them. Elsie was lying in bed, with John sitting next to her. About twenty of her friends had gathered. The place was warm from the Ashley stove, and it smelled of cornmeal and chile.

Elsie's body had shrunk, and she was barely conscious and breathing with difficulty. It seemed to me as if all the energy that five days ago had been vibrating on her surface — in her eyes, voice, and hands — was rapidly receding to a dimensionless point. I held her hand for a moment until she forcefully withdrew it, raising it in what appeared to be a signal for all of us to be still. She seemed to be straining to see someone or something. John picked her up like a little baby and cuddled her in his arms, repeatedly whispering in her ear, "Follow the light." Suddenly her old dog, sitting outside the window next to her bed, began to bay. Her physician came over and felt her pulse, listened for her heart and breath, and pronounced her dead.

I had the strangest feeling that I was witnessing a natural childbirth. The experience drastically altered my perception of death, which had been shaped by the terror I felt as a child that I would die in my sleep and burn in hell. As I wept, it seemed another poignant matter of fact that Elsie, a courageous, loving woman who had spent her life grappling with sensual realities, died in her homestead surrounded by old friends who were drinking coffee and eating posole. Things would go on as they always had. She would be buried in a grave in the apple orchard, next to her husband. The snows would melt into the Nambe River, which irrigated her trees; the willows, fleabane, and apple blossoms would come out. People would fall in love, others would feel alone, children would be born, some would die. Only the shapes would change.

ACKNOWLEDGMENTS

Elissa Melamed, peace activist, author, somatic psychologist, and artist, had an enormous impact on the shaping of this book. For more than a decade, she has constantly inspired me to make connections among various levels of human experience, particularly between the personal and the political.

Roger Guettinger, Alan and Eva Leveton, Lauren Bergen, and Susan Griffin read this manuscript at various stages and gave me invaluable feedback.

I am also grateful to Fred Hill, my agent, who persisted in curbing my impatience, insisting that I refine my ideas before rushing into the marketplace. Marie Cantlon has been an editor in the spirit of this book. Her provocative questions and criticisms helped me realize what I had set out to do. I never got the feeling that she was trying to force me to conform to her own ideas about the project.

I owe special thanks to Judith Aston, creator of Aston Patterning. Shortly after finishing *The Protean Body,* I embarked upon several months of training with her which helped me get a new perspective on the nature of somatic idealism. She also refined my understanding of how cultural objects like chairs and shoes shape our lives.

Thomas Hanna's journal, *Somatics,* has provided a forum for uniting pioneers in this field, who would otherwise be isolated because they operate outside the academic network. Without that forum, I could not have grasped the full extent of what I have called "the technology of authenticity."

I am grateful to the woman I have called Sister Ruth and Imelda for generously sharing their histories with me for inclusion in this book.

211

NOTES

BIBLIOGRAPHY

INDEX

NOTES

1. Shrinking Before Authorities

1. For a history of Dr. Rolf's development of her manipulative form of body therapy, see *Ida Rolf Talks About Rolfing and Physical Reality*, ed. Rosemary Feitis (New York: Harper & Row, 1978), and *Rolfing: The Integration of Human Structures* (New York: Harper & Row, 1979), and my book *The Protean Body* (New York: Harper & Row, 1977).

2. Reported by the Department of Health and Human Services in *The San Francisco Chronicle*, July 27, 1982, p. 2.

3. Wilhelm Reich, *The Mass Psychology of Fascism*, trans. Vincent R. Carfagno (New York: Farrar, Straus and Giroux, 1970), p. 346.

4. *The Oxford English Dictionary*, vol. II (Oxford: Clarendon Press, 1933), pp. 1011, 1012.

5. *The Oxford Latin Dictionary*, vol. II (Oxford: Clarendon Press, 1969), p. 448.

2. A Loss of Sense

1. *Bhagavad Gita*, trans. Ann Stanford (New York: Herder and Herder, 1970), pp. 22–23.

2. René Descartes, *The Meditations*, trans. L. Lafleur (New York: Liberal Arts Press, 1960), p. 84.

3. Felicia Hsin Liu, M.D., "The Crisis in Health Care," in *The San Francisco Examiner and Chronicle*, Feb. 27, 1983, *Review*, p. 9.

4. Barbara Ehrenreich and Deirdre English, *Witches, Midwives and Nurses* (Old Westbury, N.Y.: Feminist Press, 1973), p. 42.

5. Paul Starr, *The Social Transformation of American Medicine* (New York: Basic Books, 1982), p. 17.

6. Ibid., p. 32.

7. Ibid., p. 34.
8. Ehrenreich and English, *Witches*, p. 49.
9. E. Richard Brown's *Rockefeller Medicine Men: Medicine and Capitalism in America* (Berkeley: University of California Press, 1980) is an analysis of this alliance and its history.
10. An analysis and documentation of this argument appears in Thomas McKeown's *The Role of Medicine: Dream, Mirage or Nemesis* (Princeton: Princeton University Press, 1979).
11. Starr, *Social Transformation*, p. 19.
12. Brown, *Rockefeller Medicine Men*, p. 120.
13. Eva Salber, M.D., "Taking Care of Each Other," *Medical Self-Care* (Winter 1982), p. 16.
14. Brown, *Rockefeller Medicine Men*, p. 220.
15. René Dubos,, *Mirage of Health: Utopias, Progress and Biological Change* (Garden City, N.Y.: Anchor Books, 1959), p. 30, quoted by Brown.
16. *The Three Ethical Codes* (Detroit: Illustrated Medical Journal Co., 1888), p. 31, quoted by Brown, p. 66.
17. Wilhelm Reich, *The Mass Psychology of Fascism*, trans. Vincent Carfagno (New York: Farrar, Straus and Giroux, 1970), p. 346.
18. Wilhelm Reich, *The Function of the Orgasm*, trans. Vincent Carfagno (New York: Simon and Schuster, 1973), p. 360.
19. Brian Hopkins, "Culturally Determined Patterns of Handling the Human Infant," *The Journal of Human Movement Studies*, vol. 2 (1976), p. 5.
20. Ibid., p. 5.
21. Michel Foucault, *Mental Illness and Psychology*, trans. Alan Sheridan (New York: Harper & Row, 1976), p. 67.
22. Leffingwall and Robinson, *Textbook of Office Management*, quoted by Susan Griffin, *Women and Nature* (New York: Harper & Row, 1978), p. 57.
23. For an account of this process in the offices of A.T. & T., see Robert Howard, "Drugged, Bugged and Unplugged," *Mother Jones* (August 1981), pp. 39 ff.
24. Herbert Marcuse, *Negations* (Boston: Beacon Press, 1969), p. 251.

3. My Body / My Point of View

1. Melvin Konner, *The Tangled Wing: Biological Constraints on the Human Spirit* (New York: Holt, Rinehart and Winston, 1982), p. 59.
2. Henry Head, *Aphasia and Kindred Disorders of Speech* (London: Cambridge University Press, 1926), quoted in Seymour Fisher and Sidney E. Cleveland, *Body Image and Personality* (New York: Dover), 1968, p. 5.

3. "Some Points for a Comparative Study of Organic and Hysterical Paralyses," in *The Standard Edition of the Complete Psychological Works of Sigmund Freud, Vol. I*, trans. James Strachey (London: Hogarth Press, 1966), p. 169.

4. "Observations of a Severe Case of Hemi-Anaesthesia in a Hysterical Male," in *The Standard Edition*, p. 30.

5. *The Image and Appearance of the Human Body: Studies in the Constructive Energies of the Psyche* (New York: International Universities Press, 1970), p. 202.

6. Ibid., p. 211.

7. Ibid., p. 214.

8. Ibid., p. 201.

9. *Civilization and its Discontents*, trans. James Strachey (New York: W. W. Norton, 1962); *The Future of an Illusion*, trans. W. D. Robson-Scott (Garden City, N.Y.: Anchor Books, 1964).

10. *The Ego and the Id*, trans. Joan Riviere (London: Hogarth Press, 1927).

11. Wilhelm Reich, *The Function of the Orgasm*, trans. Vincent Carfagno (New York: Simon and Schuster, 1973), p. 63.

12. Ibid., p. 85.

13. Ibid., p. 145.

4. The Social Body

1. My book *The Protean Body* is devoted to analyzing various aspects of the body's plasticity. See also Peter Berger and Thomas Luckman, *The Social Construction of Reality* (Garden City, N.Y.: Anchor Books, 1967), pp. 47 ff., and Moshe Feldenkrais, *Body and Mature Behavior* (New York: International Universities Press, 1973), pp. 36 ff.

2. Marcel Mauss, "Techniques of the Body," trans. B. Brewer, *Economy and Society*, vol. 2, no. 1 (1973), pp. 70–88.

3. Margaret Mead and Frances Cooke Macgregor, *Growth and Culture: A Photographic Study of Balinese Childhood* (New York: G. P. Putnam's Sons, 1951), excerpted in *The Body Reader*, ed. Ted Polhemus (New York: Pantheon, 1978), p. 44.

4. Brian Hopkins, "Culturally Determined Patterns of Handling the Human Infant," *The Journal of Human Movement Studies*, vol. 2 (1976), p. 4.

5. Ibid., p. 12.

6. "Body Alteration and Adornment: A Pictoral Essay," in *The Body Reader*, p. 155.

7. *The San Francisco Examiner and Chronicle,* June 6, 1982, "Scene," p. 2.

8. Karl Marx, *The Economic and Philosophic Manuscripts of 1844,* trans. Martin Milligan (New York: International Publishers, 1968), p. 141.

9. Wilhelm Reich, *The Function of the Orgasm,* trans. Vincent Carfagno (New York: Simon and Schuster, 1973), p. 67.

5. Ideal Bodies

1. B. K. S. Iyengar, *Light on Yoga* (New York: Schocken Books, 1979), p. 292.

2. Ibid., p. 388.

3. The Svetasvatara Upanishad, in *The Upanishads,* trans. Juan Mascaro (Baltimore: Penguin Books, 1969). p. 88.

4. The Maitri Upanishad, ibid., p. 100.

5. Shinryu Suzuki Roshi, *Zen Mind, Beginner's Mind* (New York-Tokyo: John Weatherhill, 1980), p. 25.

6. *The San Francisco Chronicle,* Dec. 31, 1981, p. 4.

7. Ernst Krestchmer, *Physique and Character: An Investigation of the Nature of Constitution and of the Theory of Temperament,* trans. W. J. M. Sprott (New York: Humanities Press, 1951), p. 9.

8. William Sheldon, *Atlas of Men* (Darien, Conn: Hafner, 1970), p. 102.

9. M. Winckelman, quoted by Paul Schilder in *The Image and Appearance of the Human Body* (New York: International Universities Press, 1970), p. 271.

10. Thérèse Bertherat, *The Body Has Its Reasons,* trans. Carol Bernstein (Garden City, N.Y.: Avon Books, 1979), p. 92.

11. Ida Rolf, an untitled address in *The Bulletin of Structural Integration* (Sept. 1974), p. 6.

12. Stephen Jay Gould's *The Mismeasure of Man* (New York: W. W. Norton, 1981) is a marvelous history of attempts within the highest levels of science to bend scientific data to serve racism and sexism.

6. Disembodying the Enemy

1. Jonathan Schell, *The Fate of the Earth* (Garden City, N. Y.: Avon Books, 1982), p. 181.

2. Robert Scheer, *With Enough Shovels: Reagan, Bush and Nuclear War* (New York: Random House, 1982), p. 120.

3. Ibid., p. 121.

4. *Bhagavad Gita,* Ann Stanford, trans. (New York: Herder and Herder, 1970), p. 15.

5. Sigmund Freud, "Reflections on War and Death," in *Character and Culture,* ed. Philip Rieff (New York: Collier Books, 1963), p. 131.

6. *The New American Bible* (New York: P. J. Kenedy & Sons, 1970).

7. Ibid.

8. Kurt Vonnegut, *The New York Times,* June 13, 1982, p. EY 25.

9. An unpublished lecture given at the College of Santa Fe, New Mexico, October 19, 1981.

10. *The San Francisco Chronicle,* October 15, 1981, p. 5.

11. Herbert Marcuse, *Negations* (Boston: Beacon Press, 1969), p. 259.

12. *The San Francisco Chronicle,* March 9, 1983, p. 1.

13. *The New York Times,* November 1, 1981, p. 89.

14. Jack Manno, "Seizing the High Ground: The Military Race for Space," *New Age Journal* (March 1983), p. 39.

15. *The New York Times Sunday Magazine,* August 21, 1982, p. 8.

16. *The San Francisco Examiner and Chronicle,* March 14, 1982, p. A16.

17. *The San Francisco Chronicle,* October 11, 1982, p. 5.

18. *The San Francisco Examiner,* February 13, 1983, "This World" p. 3.

19. *The Wall Street Journal,* March 9, 1983, p. 18.

20. Roger Fisher, *The Graduate Review* (May/June 1981), p. 14.

21. Robert Scheer, *With Enough Shovels,* p. 124.

22. I misplaced the newspaper reference for this speech. When I telephoned Mr. Turner's office, I was informed that he never makes notes of speeches or keeps clippings.

7. A Woman's Way

1. A. Peerbhai, quoted by Leora Rosenthal, "The Definition of Female Sexuality and the Status of Women among the Gujerati-Speaking Indians of Johannesburg," in John Blacking, ed., *The Anthropology of the Body* (New York: Academic Press, 1977), p. 192.

2. Katherine Arnold, "The Introduction of Poses to a Peruvian Brothel and Changing Images of Male and Female," in Blacking, *Anthropology of the Body,* p. 192.

3. Ibid., p. 180.

4. *The New American Bible,* (New York: P. J. Kenedy & Sons, 1970).

5. *The Penitential of Theodore,* in John T. McNeill and Helena M. Gamer, *Medieval Handbooks of Penance* (New York: Octagon Books, 1965), XIV, 17.

6. Seymour J. Gray, M.D., "Beyond the Veil," *The San Francisco Chronicle,* March 3, 1983, p. 40.

7. Dorothy Dinnerstein, *The Mermaid and the Minotaur: Sexual Arrangements and Human Malaise* (New York: Harper & Row, 1977), p. 132.

8. Ibid., p. 133.

9. Sigmund Freud, *Civilization and its Discontents,* trans. James Strachey, (New York: W. W. Norton, 1962), p. 50.

10. Quoted by Wilhelm Reich in *The Mass Psychology of Fascism,* trans. Vincent Carfagno (New York: Farrar, Straus and Giroux, 1970), p. 61.

11. Quoted in an undated report of People for The American Way, 1015 18th Street, N.W., Suite 300, Washington, D.C. 20036.

12. Ibid.

13. Dinnerstein, *Mermaid and Minotaur,* p. 13.

14. Ibid., p. 153.

15. Frederick Leboyer, *Loving Hands: The Traditional Indian Art of Baby Massaging* (New York: Knopf, 1976).

16. Brian Hopkins, "Culturally Determined Patterns of Handling the Human Infant," *The Journal of Human Movement Studies,* vol. 2 (1976), pp. 10–12.

17. Barbara Ehrenreich and Deirdre English, *Witches, Midwives and Nurses* (Old Westbury, N. Y.: Feminist Press, 1973), pp. 30, 33.

18. Cited by Mary Daly, *Gyn/Ecology* (Boston: Beacon Press, 1978), p. 183 note.

19. Ehrenreich and English, *Witches,* p. 35.

20. Ibid.

21. E. Richard Brown, *Rockefeller Medicine Men* (Berkeley: University of California Press, 1980), p. 92.

22. Michel Foucault, *The History of Sexuality, Vol. I,* trans. Robert Hurley (New York: Vintage Books, 1980), p. 55 and note.

23. Barbara Ehrenreich and Deirdre English, *For Her Own Good: 150 Years of the Experts' Advice to Women* (Garden City, N.Y.: Anchor Press, 1979), p. 97.

24. Elsa Gindler, "Gymnastik for Busy People," an unpublished

manuscript trans. from the German, "Die Gymnastik des Berufsmenschen," originally published in 1926 in *Gymnastik,* the journal of the German Gymnastik Foundation.

8. Coming to Our Senses

1. Ilana Rubenfeld, "An interview with Charlotte Selver and Charles Brooks," *Somatics* (Spring 1977), p. 19.

2. Nikolas Tinbergen, "Ethology and Stress Diseases," *Science,* vol. 185 (July 5, 1974), p. 20–27.

3. "The Australian Story," in *The Resurrection of the Body,* ed. Edward Maisel (New York: Dell, 1969), p. 161.

4. "About Golf," ibid., pp. 120, 121.

5. "After the Bomb," ibid., p. 89.

6. "About Golf," ibid., p. 118.

7. "The Trouble with Physical Exercises," ibid., p. 110.

8. Ilana Rubenfeld, ibid., p. 15.

9. Ibid., p. 14.

10. Charles Brooks, *Sensory Awareness: The Rediscovery of Experiencing* (New York: Viking, 1974), p. 7.

11. Moshe Feldenkrais, *The Case of Nora* (New York: Harper & Row, 1977), p. 13.

12. Ibid., p. 44.

13. From an unpublished communication.

14. John Naisbitt, *Megatrends: Ten New Directions Transforming Our Lives* (New York: Warner Books, 1982), p. 198.

15. Moshe Feldenkrais, *Body and Mature Behavior: A Study of Anxiety, Sex, Gravitation and Learning* (New York: International Universities Press, 1973), p. 13.

9. Consensus

1. Norman O. Brown, *Love's Body* (New York: Vintage Books, 1968), p. 82.

2. Boston Women's Health Collective, *Our Bodies, Ourselves* (New York: Simon and Schuster, 1979).

3. Krieger describes her work in *The Therapeutic Touch: How to Use Your Hands to Help or Heal* (Englewood Cliffs, N.J.: Prentice-Hall, 1979).

4. One such example is described in Hugh V. Delaney, "Teaching Medicine in Cambodia," *Medical Self-Care* (Winter 1982), p. 22.

5. Marlow Hotchkiss, "The Mo-Tzu Project," *CoEvolution Quarterly* (Fall 1982), p. 84.

6. John Vasconcellos, *A Liberating Vision: Politics for Growing Humans* (San Luis Obispo: Impact Press, 1979).

7. See, for example, Will Schutz's *The Human Element* (Berkeley: Ten Speed Press, 1983).

8. Paul Hawken, "Illusory Inflation," *CoEvolution Quarterly* (Summer 1981), p. 37, quoting George Land, *Grow or Die* (New York: Dell, 1974).

9. Sir Herbert Read, "The Biological Significance of Art," *The Saturday Evening Post*, Sept. 26, 1959; quoted by Roderyk Lange, "Some Notes on the Anthropology of Dance," in John Blacking, ed., *The Anthropology of the Body*, (New York: Academic Press, 1977), p. 250.

10. Timothy Gallwey's *The Inner Game of Tennis* (New York: Random House, 1974), and Gallwey and Bob Kriegel's *The Inner Game of Skiing* (New York: Random House, 1977), describe in more detail methods for teaching sports along these lines.

10. Consensual Spirituality

1. Herbert Thurston, *The Physical Phenomenon of Mysticism* (Chicago: Regnery, 1952).

2. Ibid., p. 198.

3. Friederich Heer, *The Intellectual History of Europe*, 2 vols., trans. J. Steenberg (New York: Doubleday, 1968); George Williams, *The Radical Reformation* (Philadelphia: Westminster Press, 1962); Ronald Knox, *Enthusiasm* (New York: Oxford University Press, 1950).

4. Jeremiah 31:31; John 6:45.

5. Norman O. Brown, *Love's Body* (New York: Vintage, 1968), p. 82.

BIBLIOGRAPHY

Alexander, F. Matthias. *The Resurrection of the Body.* New York: Dell, 1969.

Argyle, Michael. *Bodily Communication.* New York: International Universities Press, 1975.

Bateson, Gregory. *Mind and Nature: A Necessary Unity.* New York: Bantam, 1980.

———. *Steps to an Ecology of Mind.* New York: Ballantine, 1972.

Berthérat, Thérèse. *The Body has its Reasons.* Trans. Carol Bernstein. New York: Avon, 1979.

Blacking, John, ed. *The Anthropology of the Body.* New York: Academic Press, 1977.

Blechschmidt, Erich. *The Beginnings of Human Life.* New York: Springer, 1977.

Bloomer, Kent. *Body, Memory and Architecture.* New Haven: Yale University Press, 1977.

Boston Women's Health Collective. *Our Bodies, Ourselves.* New York: Simon and Schuster, 1979.

Bourguignon, Erika, ed. *Religion, Altered States of Consciousness and Social Change.* Columbus, Ohio: University of Ohio Press, 1973.

Brooks, Charles. *Sensory Awareness: The Rediscovery of Experiencing.* New York: Viking, 1974.

Brown, E. Richard. *Rockefeller Medicine Men: Medicine and Capitalism in America.* Berkeley: University of California Press, 1980.

Brown, Norman O. "Apocalypse: The Place of Mystery in the Life of the Mind." *Harper's* 222 (May 1961): 46–49.

_____. *Closing Time*. New York: Random House, 1973.

_____. "Daphne or Metamorphosis." *Myths, Dreams and Religion*. Ed. Joseph Campbell. New York: Dutton, 1970.

_____. *Life Against Death: The Psychoanalytical Meaning of History*. New York: Vintage/Random House, 1968.

_____. *Love's Body*. New York: Random House, 1966; Vintage, 1968.

_____. "From Politics to Metapolitics." *Caterpillar* 1 (October 1967): 62–94.

_____. "A Reply to Herbert Marcuse." *Commentary* 43 (March 1967): 83–84.

Buytendijk, F. J. J. *Attitude et mouvements: Étude fonctionelle du mouvement humaine*. Paris: Desclée de Brouwer, 1957.

Cannon, Walter B. *The Wisdom of the Body*. New York: W. W. Norton, 1963.

Chernin, Kim. *The Obsession: Reflections on the Tyranny of Slenderness*. New York: Harper & Row, 1981.

Daly, Mary. *Gyn/Ecology*. Boston: Beacon Press, 1978.

Darwin, Charles. *The Expression of Emotion in Man and the Animals*. Chicago: University of Chicago Press, 1965.

Davis, Martha. *Towards Understanding the Intrinsic in Body Movement*. New York: Arno Press, 1975.

de Beauvoir, Simone. *The Second Sex*. Trans. H. M. Parshley. Vintage/Random House, 1974.

Descartes, René. *Discourse on Method* and *Meditations*. Trans. L. Lafleur. New York: Liberal Arts Press, 1960.

Dinnerstein, Dorothy. *The Mermaid and the Minotaur: Sexual Arrangements and Human Malaise*. New York: Harper & Row, 1977.

Douglas, Mary. "Do Dogs Laugh?" *Journal of Psychosomatic Research* 15 (1971): 387–390.

_____. *Natural Symbols: Explorations in Cosmology*. London: Barrie and Rockliff, 1970.

_____. *Purity and Danger: An Analysis of Concepts of Pollution and Taboo*. New York: Praeger, 1966.

_____. *Rules and Meanings: The Anthropology of Everyday Knowledge*. London: Penguin, 1973.

Duncan, Isadora. *My Life*. New York: Liveright, 1955.

Ehrenreich, Barbara, and English, Deirdre. *Complaints and Dis-

orders: The Sexual Politics of Sickness. Old Westbury, N.Y.: Feminist Press, 1973.

_____. *For Her Own Good: 150 Years of the Experts' Advice to Women.* Garden City, N.Y.: Anchor/Doubleday, 1979.

_____. *Witches, Midwives and Nurses.* Old Wesbury, N.Y.: Feminist Press, 1973.

Englehardt, H. Tristram, Jr. "Bioethics and the Process of Embodiment." *Perspectives in Biology and Medicine* 18, no. 4 (Summer 1975): 486–500.

_____. *Mind-Body: A Categorical Relation.* The Hague: Martinus Nijhoff, 1973.

Feldenkrais, Moshe. *Awareness Through Movement: Health Exercises for Personal Growth.* New York: Harper & Row, 1972.

_____. *Body and Mature Behavior: A Study of Anxiety, Sex, Gravitation and Learning.* New York: International Universities Press, 1970.

_____. *The Case of Nora: Body Awareness as Healing Therapy.* New York: Harper & Row, 1977.

Fenton, J. Y., ed. *Theology and the Body.* Philadelphia: Westminster Press, 1974.

Firth, R. F. *Symbols Public and Private.* London: Allen and Unwin, 1973.

Fisher, Seymour, and Cleveland, Sidney. *Body Image and Personality.* New York: Dover, 1968.

Foucault, Michel. *The Birth of the Clinic: An Archaeology of Medical Perception.* Trans. A. Smith. New York: Vintage/Random House, 1975.

_____. *The History of Sexuality. Volume I: An Introduction.* Trans. Robert Hurley. New York: Vintage/Random House, 1978.

_____. *Mental Illness and Psychology.* Trans. Alan Sheridan. New York: Harper & Row, 1976.

_____. *The Order of Things: An Archaeology of the Human Sciences.* New York: Vintage/Random House, 1970.

_____. *Power/Knowledge: Selected Interviews and Other Writings, 1972-1977.* Ed. Colin Gordon. New York: Pantheon, 1980.

Franklin, K. J. "Kewa Ethnolinguistic Concepts of Body Parts." *Southwestern Journal of Anthropology* 19, no. 1: 54–63.

Freud, Sigmund. *Character and Culture.* Ed. Philip Rieff. New York: Macmillan, 1963.

_____. *Civilization and Its Discontents.* Trans. James Strachey. New York: W. W. Norton, 1962.

_____. *The Future of an Illusion.* Trans. W. D. Robson-Scott. New York: Doubleday/Anchor, 1961.

_____. "Observations of a Severe Case of Hemi-Anaesthesia in a Hysterical Male." Trans. James Strachey. *The Standard Edition, Vol. I.* London: Hogarth, 1966, pp. 21–31.

_____. "Psychical (or Mental) Treatment." Trans. James Strachey. *The Standard Edition, Vol. VII.* London: Hogarth, 1953, pp. 281–302.

_____. "Some Points for a Comparative Study of Organic and Hysterical Paralyses." Trans. James Strachey. *The Standard Edition, Vol. I.* London: Hogarth, 1966, pp. 155–172.

_____. *Totem and Taboo.* Trans. A. A. Brill. New York: Vintage/Random House, 1946.

Gallwey, Timothy. *The Inner Game of Tennis.* New York: Random House, 1974.

_____, and Kriegel, Bob. *The Inner Game of Skiing.* New York: Random House, 1977.

Gindler, Elsa. "Gymnastik for Busy People." An unpublished translation of "Die Gymnastik des Berufsmenschen," which appeared in the 1926 edition of *Gymnastik,* the journal of the German Gymnastik Foundation.

Gould, Stephen Jay. *The Mismeasure of Man.* New York: W. W. Norton, 1981.

_____. *The Panda's Thumb.* New York: W. W. Norton, 1980.

Greer, Germaine. *The Female Eunuch.* London: Granada, 1970.

Griffin, Susan. *Woman and Nature.* New York: Harper, 1978.

Grossinger, Richard. *Planet Medicine: From Stone Age Shamanism to Post-Industrial Healing.* Boulder, Colo.: Shambhala, 1982.

Hallpike, C. R. "Social Hair." *Man* 4, no. 2: 256–267.

Hanna, Thomas. *Bodies in Revolt.* New York: Holt, Rinehart and Winston, 1970.

_____. *The Body of Life.* New York: Knopf, 1979.

Head, Sir Henry. *Aphasia and Kindred Disorders of Speech.* London: Cambridge University Press, 1926.

_____. "On Disturbances of Sensation with Especial Reference to the Pain of Visceral Disease." *Brain* 16 (1893): 1 ff.

_____. *Studies in Neurology.* 2 vols. London: Cambridge University Press, 1920.

Heer, Friederich. *The Intellectual History of Europe.* 2 vols. Trans. J. Steenberg. New York: Doubleday, 1968.

Hertz, Robert. *Death and the Right Hand.* Trans. R. and C. Needham. Aberdeen: Cohen and West, 1960.

Hewes, Gordon. "The Anthropology of Posture." *Scientific American* 196 (February 1957): 123-132.

──────. "World Distribution of Certain Postural Habits." *American Anthropologist* 57 (1955): 231-244.

Hinde, R. A. *Biological Bases of Human Social Behavior.* New York: McGraw-Hill, 1974.

──────. *Non-Verbal Communication.* London: Cambridge University Press, 1972.

Hopkins, Brian. "Culturally Determined Patterns of Handling the Human Infant." *The Journal of Human Movement Studies* 2 (1976): 1-27.

Johnson, Don. "The Body: The Cathedral and the Kiva." *In Search of a Therapy.* Ed. Dennis Jaffe. New York: Harper & Row, 1975, pp. 139-150.

──────. *The Protean Body.* New York: Harper & Row, 1977.

──────. "Somatic Platonism." *Somatics* 3, no. 1 (Autumn 1980): 4-7.

Jonas, Hans. *The Phenomenon of Life: Toward a Philosophical Biology.* New York: Delta, 1966.

Katz, Michael. *Class, Bureaucracy and Schools: The Illusion of Educational Change in America.* New York: Praeger, 1971.

Kern, S. *Anatomy and Destiny: A Cultural History of the Human Body.* New York: Bobbs-Merrill, 1975.

Knox, Ronald. *Enthusiasm.* New York: Oxford University Press, 1950.

Konner, Melvin. *The Tangled Wing: Biological Constraints on the Human Spirit.* New York: Holt, Rinehart and Winston, 1982.

Kretschmer, Ernst. *Physique and Character: An Investigation of the Nature of Constitution and of the Theory of Temperament.* Trans. W. Sprott. New York: Humanities Press, 1951.

Krieger, Dolores. *The Therapeutic Touch: How to Use Your Hands to Help or Heal.* Englewood Cliffs, N.J.: Prentice-Hall, 1979.

Kunzle, David. *Fashion and Fetishism: A Social History of the Corset, Tight-lacing, and Other Forms of Body-Sculpture in the West.* London: Rowman and Littlefield, 1981.

Leboyer, Frederick. *Loving Hands: The Traditional Indian Art of Baby Massaging.* New York: Knopf, 1976.

Lewis, Gilbert. "Gnau Anatomy and Vocabulary for Illnesses." *Oceania* 45 (1974): 50-78.

228 / BODY

Mahr, August C. "Anatomical Terminology of the Eighteenth Century Delaware Indians: A Study in Semantics." *Anthropological Linguistics* 2 (1960): 1–65.

Marcuse, Herbert. *Eros and Civilization.* Boston: Beacon Press, 1955.

_____. *Negations.* Boston: Beacon Press, 1968.

_____. *One-Dimensional Man.* Boston: Beacon Press, 1964.

Marsh, G. H., and Laughlin, W. S. "Human Anatomical Knowledge Among the Aleutian Islanders." *Southwestern Journal of Anthropology* 12 (1956): 38–78.

Marx, Karl. *Capital.* Trans. S. Moore and E. Aveling. New York: Modern Library (n.d.).

_____. *The Economic and Philosophic Manuscripts of 1844.* Trans. Martin Milligan. New York: International Publishers, 1964.

_____. *The German Ideology.* Ed. R. Pascal. New York: International Publishers, 1968.

Mauss, Marcel. "Éffect physique chez l'individu de l'idée de mort suggerée par la collectivité." *Journal de psychologie normale et pathologique* 23: 653–669.

_____. "Les techniques du corps." *Journal de psychologie normale et pathologique* 32 (1935). Trans. B. Brewer; "The Techniques of the Body." *Economy and Society* 2 (1973): 70–88.

McCulloch, Warren. *Embodiments of Mind.* Boston: MIT Press, 1965.

McKeown, Thomas. *The Role of Medicine: Dream, Mirage or Nemesis?* Princeton, N.J.: Princeton University Press, 1979.

McNeill, William H. *The Pursuit of Power: Technology, Armed Force, and Society since A.D. 1000.* Chicago: University of Chicago Press, 1965.

Melamed, Elissa. *Mirror, Mirror: The Terror of Not Being Young.* New York: Linden/Simon and Schuster, 1983.

Merleau-Ponty, Maurice. *The Phenomenology of Perception.* Trans. Colin Smith. London: Routledge and Kegan Paul, 1962.

_____. *The Primacy of Perception.* Ed. James Edie. Chicago: Northwestern University Press, 1964.

_____. *The Structure of Behavior.* Trans. Alden Fisher. Boston: Beacon Press, 1963.

Needham, R. *Right and Left: A Study in Religious Polarity.* Chicago: University of Chicago Press, 1974.

Perey, A. "Body and World in Oksapmin Kin Terms." *Oceania* 45 (1975): 235–236.

Pietsch, Paul. *Shufflebrain: The Quest for the Hologramic Mind.* Boston: Houghton Mifflin, 1981.

Polhemus, Ted, ed. *The Body Reader: Social Aspects of the Human Body.* New York: Pantheon, 1978.

_____, and Benthall, Jonathan, eds. *The Body as a Medium of Expression.* New York: Dutton, 1975.

Pribram, Karl, ed. *Brain and Behavior.* New York: Penguin, 1969.

Prigogine, Ilya. *From Becoming to Being.* San Francisco: W. H. Freedman, 1980.

Reich, Wilhelm. *The Function of the Orgasm.* Trans. Vincent Carfagno. New York: Simon and Schuster, 1973.

_____. *The Mass Psychology of Fascism.* Trans. Vincent Carfagno. New York: Farrar, Straus and Giroux, 1970.

Rieber, R. W., ed. *Body and Mind: Past, Present and Future.* New York: Academic Press, 1980.

Rolf, Ida. *Ida Rolf Talks about Rolfing and Physical Reality.* Ed. Rosemary Feitis. New York: Harper & Row, 1978.

_____. *Rolfing: The Integration of Human Structures.* Santa Monica, Calif.: Dennis-Landman, 1977.

Rubenfeld, Ilana. "An Interview with Charlotte Selver and Charles Brooks." *Somatics* 1 (Spring 1977): 14–20.

Sayres, William T. *Body, Soul and Blood: Recovering the Human in Medicine.* Troy, Mich.: Asclepiad Publications, 1980.

Scheer, Robert. *With Enough Shovels: Reagan, Bush and Nuclear War.* New York: Random House, 1982.

Schell, Jonathan. *The Fate of the Earth.* New York: Avon, 1982.

Schilder, Paul. *The Image and Appearance of the Human Body: Studies in the Constructive Energies of the Psyche.* New York: International Universities Press, 1970.

Schutz, Will. *The Human Element.* Berkeley: Ten Speed Press, 1983.

Sheldon, W. H. *Atlas of Men.* Darien, Conn: Hafner, 1970.

_____. *The Varieties of Temperament: A Psychology of Constitutional Differences.* New York: Harper Bros., 1942.

Singer, C. *A Short History of Anatomy and Physiology from the Greeks to Harvey.* New York: Dover, 1957.

Spicker, S. F., ed. *The Philosophy of the Body.* Chicago: Quadrangle, 1970.

Spring, Joel. *Education and the Rise of the Corporate State.* Boston: Beacon Press, 1972.

Starr, Paul. *The Social Transformation of American Medicine: The*

Rise of a Sovereign Profession and the Making of a Vast Industry. New York: Basic Books, 1982.

Straus, Erwin. *Phenomenological Psychology: Selected Papers.* Trans. E. Eng. New York: Basic Books, 1966.

Suzuki, Shinryu Roshi. *Zen Mind, Beginner's Mind.* New York: John Weatherhill, 1980.

Tinbergen, Nikolas. "Ethology and Stress Diseases." *Science* 185 (July 5, 1974): 20–27.

Todd, Mabel Elsworth. *The Thinking Body.* Brooklyn: Dance Horizons, 1937.

Ulil, Michael, and Ensign, Tod. G. I. *Guinea Pigs: How the Pentagon Exposed Our Troops to Dangers More Deadly Than War.* Chicago: Playboy Press, 1980.

VanDenBerg, J. H. "The Human Body and the Significance of Human Movement." *Philosophy and Phenomenological Research* 13 (December 1952): 159–183.

Wood, Corinne Shear. *Human Sickness and Health: A Biocultural View.* Palo Alto, Calif.: Mayfield, 1979.

Werner, O., and Begishe, K. Y. "A Lexemic Typology of Navajo Anatomical Terms I: The Foot." *International Journal of American Linguistics* 36 (1970): 247–265.

Williams, George. *The Radical Reformation.* Philadelphia: Westminster Press, 1962.

Zaner, Richard. *The Context of Self: A Phenomenological Inquiry Using Medicine as a Clue.* Athens: Ohio University Press, 1981.

INDEX

Don Johnson, author of *The Protean Body,* works directly with people's bodies as a therapist in private practice. He also directs a clinical training program in somatic studies and is a member of the graduate faculty at Antioch University West in San Francisco.